LOCH NESS

TALES LEGENDS AND RECIPES

From the Highlands of Scotland and beyond.

By

Hugh Fraser.

Dedicated to the memory of

Valerie Whitehead, a good friend and fellow cruising sailor.

Many thanks to Brenda, for putting up with me during the
hundreds of hours this book took to write.

To see pictures of Loch Ness

Go to <www.tinkerbell-images.co.uk>

The short stories are underlined. All sketches drawn by the author.

7 A **Night Out**. An encounter with the Loch Ness Monster.

17 The Loch Ness Monster or 'Nessie and the first recorded sighting.

18 Urquhart Castle. Sketch and a few words about the castle.

19 Lammas Time Highland cattle and the drovers.

19 Crowdie. Traditional Scottish cheese.

20 Oatcakes, modern and ancient.

21 **In the Village Inn**. Yarn about people who stayed by the loch.

31 Straw monster.

32John Cobb

32 Murder at Glen Coe. About the crime.

33 Balled of Glen Coe.

34 Vitrified forts. A little about the ancient fort above Loch Ness.

34 Blaeberry jam. The half way inn

35 The half way Inn. Robbery at the Loch side inn.

35 Clootie Dumpling.

36 **The Grouse and Claret**. An evening's trout fishing.

44 The phantom hand. Dores, Loch Ness. Jacobite graves.

45 Loch Ness trout.

46 **The Card**. A tale of first love.

50 The Romans visit to Scotland. How the Celts welcomed them

51 Macaroon bars.

52 **A Man for the Drink**. A man and his whisky. Sketch.

61 Origins of whisky. Athol Brose, hot toddy, whiskey butter.

63 Oatmeal Bannocks.

64 **The Exile**. Sketch, Bush Camp. About the Australian gold fields. Recipe just for fun.

71 Host a Burns Supper.

72. Words, To a Haggis.

73 Recipes. Haggis, old and new and rabbit hotpot.

75 The Bell. Moving story about a full life and a final homecoming.

83 The Gondolier. A steamer that once sailed on Loch Ness.

84 Scapa Flow Orkney. A little about her final resting place.

85 Hogmanay New Year Bad' (Ball) Game.

86 Black bun. First footing.

87 One for the Pot. A yarn about a poacher. Sketch.

96 The Lovat Scouts. A few words about this famous Company.

96 Mealy Pudding. Another ancient oat based meal.

97 Ancient sweet oat biscuits, (cookies.)

97 Tattie Soup.

98 Cock o' Leekie Soup.

98 **The Huntress**. A Highland vixen and her cubs.

103 Clapshot.Minced Collops.

104 **The Wee Lairdie**. Life long friendship. Become a Scottish Laird

111 Shortbread. Two recipes.

112 **The Statement**. Yarn about a missing day

117 A few famous Scots.

119 Cullen Skink and salmon fish cakes.

120 The Stone of Destiny.

121 **Sandy's Widow**. A widow who returns to her village.

129 Stovies. Traditional Scottish meal.

130 Kishmul's' Galley, song from the Isle of Barra.

131 Scottish Trifles.

131 **The White Horse Treasure**. A murderous night's work.

142 The Culbin sands, a little about the Highland desert.

143 History, in songs.

154 The Highland Clearances. A heinous crime.

155 **The Keeper**. Yarn about a Highland gamekeeper.

161 **An ear for the unheard**.

167 Second Sight, the Brahn Seer.

177 Ancient Pagan festivals. Witches etc

169 **The Salmon**. A salmon returns to Loch Ness. Recipe.

173 Loch Ness Fisherman in a time of plenty. Photo.

175 A few people from the parish of Dores 1950's.

175 **The Object**. Science fiction, Highland style.

185 Ring stories.

185 Mutton Pies. The 'dinner' of many a working man.

186 **The End.** Story about the consequences of climate change.

187 Scottish words used in this book.

189 About the Author.

Other book by the Hugh Fraser

A Night out on Loch Ness.

If I kinda squint, I can almost see him now. He would be standing before me in his green, well-worn wellies. A black grubby boiler suit would hang loosely from his slight frame and a threadbare fore-and-aft hat perched, on the back of his head. He would look up at me, as a lippy toothless grin spread over his tanned weather-beaten face. His bushy eyebrows would arch, as a mischievous twinkle appeared in his sea blue eyes. His hands would be working away making one of his anorexic roll-up-cigarettes. His eyes would never be still, for they would dart this way and that, peering, searching, then occasionally pausing for a moment, on some object or other. When he was satisfied he had seen everything in his visual range, he would return his eyes to his cigarette and examine it for a moment. Then with a deft practiced movement, he would lick the glue with the tip of his tongue, pop the whole thing into his mouth and draw it out between his lips, trapping it at the very last moment. A match would flare in his cupped hands followed shortly after, by his usual cough.

Aye, that was Duncan all right, and I suppose he was a fine man in his own way. Although I seem to recall he was a bit on the wild side in his youth.

It must be about five years ago, that he called to see me. Man I remember the day well; as if it was only yesterday.

'**Duncan**! And how are you today?' I asked as I came out of my loch side cottage and saw him leaning against his old van.

'Och you're seeing it,' he replied, using his usual retort as he carefully examined the end of his newly manufactured smoking creation. His face contorted for a moment as his fleshy lips worked away at some hidden task. Then he turned his head to the side and spat out an almost invisible strand of tobacco.

'And yourself, Bob,' he continued after the short pause.'

'Oh I'm just fine, just fine,' I replied as I stood waiting for him to tell me what he wanted.

'And the bee-ssss,' he continued glancing at my garden hives, 'they're doing fine too I take it. Man it's a grand fine spell of settled weather were having, and I'm thinking the heather might be out a wee bit early this year,' he added as he cast his eyes over the loch to the far hillside.

'Oh they're doing fine too,' I replied, 'just fine.'

'I was just, kinda wondering,' he said five minutes later, as his patter began to wane. He shuffled his feet and looked around. Then he wiped his mouth and his bristly chin with his hand. I thought here it comes. 'If you might just happen to have a wee bit o' net, kinda in your way. Just an old bit you'll no be needin. No matter if it's got a few holes in it.'

'It would be no use to you Duncan if it had no holes in it,' I interrupted in reply.

'Och just so Bob, just so. You're a fair comic man, a fair comic. It's the stage you should be on, people would pay good money to hear humour like yon.'

Ignoring his sarcasm, I continued, 'so you're planning another go at the salmon Duncan.'

'Wheesht man, who told you that, I was just after a bit of net to keep the birds off my rasps; they're fair making a meal o' them.'

'Rasps Duncan! I never knew you had rasps.'

'Oh I have rasps all right, lots o' them.' The lie leaving his lips as easy as breathing. 'You'll no have a spare bit of net about the place then?'

'Oh I might, I might,' I replied wondering how much I could get for it, 'but I was going to swap it with a man I know for a bit of fish. And as you only want it for your rasps, well, I might be better hanging on to it for now.'

'It's a bit of fish you want!' Duncan exclaimed, 'well, why did you no just say so? I know a man who has lots o' fish. I'll drop you off a nice one, next time I'm passing'.

'I was perhaps thinking the net might be worth four or five fish,' I suggested.

'Och you're a hard man right enough Bob,' he said looking around as he rubbed his chin again. 'Tell you what,' he said, as he enacted the hand spitting ritual, 'I'll give you two fish for-it'. I followed suit and held out my hand.

'Make it three.'

'Done,' he said and we slapped hands on the deal.

'Just the one wee thing,' I said, as I returned from my shed with the net, 'if you get yourself caught wi' this, it never came from me.'

'Just so,' he said nodding his head, 'I'll just say I found it. Same as I did last time.' Then Duncan, who now had what he wanted, wandered across and stuffed the old net into his even older van. He drove away without a backwards glance, leaving a cloud of blue exhaust smoke in his wake.

Och well I thought, that saves me the bother of carrying out the wife's orders to burn it.

My wife, the newly elected president of the village W.R.I. (Women's Rural Institute,) felt rather strongly, that my wee bit of poaching was not compatible with her new social standing, or her lofty new office.

A week or two later, Duncan drove around the back of my cottage. I watched from my shed, as he got out of his van and went through his cigarette ritual again.

'Well Duncan,' I said when he spotted me, 'you've had a busy night no doubt.'

'Wheesht man, no so loud,' he replied looking around. 'I've a wee something in the van for you.' He rubbed his ear and looked around again.' It's in sort of payment,' he added quietly. We walked over to his van, and when he opened the back door, I saw a haunch of venison lying on an old sack.

'What's this Duncan? I kinda remember it was three salmon we agreed on.'

'Just so,' he replied in a quiet voice, 'but you see I never quite got around to fixing yon bit of net, and I happened on this venison wandering on the road last night. Lost it was. So I thought you might like a nice bit o' meat instead.'

'Well that's very good of you Duncan, but I've a freeze full of the stuff yonder and the wife is really fond of the fish. Och well,' I said shaking my head, 'since you have it with you, and it's no safe driving around with the Laird's venison, you can leave it here. But I'll still be wanting my fish.'

Duncan spat a pretend spit, on his hand, 'so it's just the one fish I'm due you then.'

'Two fish,' I replied holding my hand out of range.

'Done' and as we slapped hands on our deal, I became the unlawful owner of a nice haunch of poached venison.

When I gave it to my wife, she was delighted. 'Venison!' she cried, 'we've no had venison for months.'

Several weeks passed by and I thought Duncan had forgotten his debt. Then one afternoon I looked up from my vegetable garden and saw him leaning on a tree watching me.

'You'll have my fish with you, Duncan?' I asked in way of a greeting.

'Man Bob, yon net took some fixing, it only had half the holes in it, it was supposed to have; it took me ages to mend.'

'You'll have my fish then?' I asked again ignoring his gripe.

'Just so, they're in the van.' We wandered over to his van; him having a good look around as he went. He removed a sack from the back and handed it to me. I opened the bag and saw two nice fresh run salmon, both about ten pounds in weight.

'I'll be needing the bag back.' He called as I took the fish inside. I returned a few minutes later and tossed the sack into his van. Duncan slammed the door shut, and then leaned back. He lifted his head and scrutinised me carefully. What now I thought, as he rubbed his lips and then his bristly chin. 'I was just kinda wondering,' he said looking around as he transferred his hand to the back of his neck. If you'd be, interested in a wee night out. I've kinda got the lend-o' this boat, and I know a bay that's fair hoaching with fish. I thought, maybe you might want to give me a wee hand.'

'Och Duncan I would love to, but I promised the wife I would leave the poaching alone. Since you hit the headlines in the Inverness Courier, she's been kinda concerned, that like you, I might get my name in the paper and disgrace her in front of all her new friends.'

'Just so,' he replied shaking his head. 'It's a terrible responsibility she's taken on, what with the Lairds wife, and the Doctors wife, and that councillor wifie attending the meetings.'

'Och well,' I replied, 'it makes her happy.'

'You'll no doubt have heard the keeper's son's getting himself married.

'No! 'I replied, 'who to?

Duncan's weather beaten face changed into a broad lippy grin. 'Oh she's a bonny lassie right enough,' he said with his eyes sparkling. Then he leaned forward and whispered, 'she's the daughter of yon water bailiff mannie.'

'And just when is this wedding to take place, Duncan?' I asked, trying not to show interest. Duncan slowly drew his fathers old watch out of his pocket, he slowly opened the lid and squinted at the dial.

'In about four hours' time,' he said looking up into my eyes. 'The reception is up at the Drumossie hotel,' he continued, 'so they'll no be home to night.'

Now I rubbed my mouth and then my chin. 'Tempting,' I replied, 'but are you quite sure? They've kept it kinda quiet.'

'Oh yes,' he said with a lippy grin.

'You know Willie Shaw?

'Him that plays the fiddle?' I asked.

'Aye that's him; I met him in the Inn last night. He told me he was booked to play at the wedding. You know in the ceilidh band. Well he said the hotel had an extra lorry o' whisky delivered.'

'So they'll definitely not be home tonight.' I replied realising that I could not fault his logic.

'Just so,' Duncan muttered nodding his head. I hummed and hawed for a few minutes but the information was just too good to ignore, so I accepted his offer of a night out.

Duncan then asked me if I could collect him, as his van was not working too well.

'Oh what a pity you never said so earlier,' I lied, 'my wife is going to visit her sister in Nairn tonight, and she'll more than likely stay over.'

'Ah well I just thought I'd ask,' he replied in a disappointed voice as he stared at me through half closed eyes, 'I dare say my old van will get us there.' I had no intention of putting my good car at risk. If we got caught, Anderson the Magistrate, would take it from me. Duncan agreed to collect me at midnight and as he drove away, I went inside with the difficult task of convincing my wife that the night out carried very little risk. I told her about the wedding, taking care to point out that every keeper and bailiff for fifty miles around would be at the reception, and that we could fill the freezer with salmon. Eventually I convinced her with the solemn promise that this would be the last time I would ever go poaching.

Duncan arrived just as the clock struck the witching hour. I stepped out of the shadows and after looking around slid into the passenger seat of his old van.

'Where is this bay?' I asked as we drove along the narrow road by the Loch side.

'Oh it's a fair bit away. It's just over the other side of the loch.'

'Across the other side! You want me to cross Loch Ness in the dark! In a wee boat?'

'We'll have an engine, so you'll no have to row,' he assured me.

'But the Beastie,' I said with real concern. 'Everyone knows she comes out at night.'

'Dinna be daft man, that's just a wee story for the tourists.' I was far from convinced but as I was somewhat committed to the night out, I put my fears aside and sat trying not to think about the Loch Ness Monster.

Half an hour later, Duncan parked his van in the loch-side hazel wood, and we sat quietly for ten minutes. Just to be sure, we were alone. Then as we crept down the wooded bank to the boathouse, a larch clinker built boat came into view. It looked a bit on the wee side, and a bit auld-warldie, (ancient.) I doubted if the near derelict was 12 foot long. As I examined its peeling paint, and ran my hand under its bilges checking for holes, a half moon crept out from behind a cloud. Suddenly the menacing waters of Loch Ness stretched out before me. I swallowed hard and dumped the net in the boat. I tore my eyes away and saw Duncan, who was making a show of looking for the shed key.

'Man, he said it was just under this stone, or perhaps it was this one. Och well not to worry. I'll just burst the lock. He'll no mind; in fact he would prefer it.' With that, he produced a big screwdriver, which he just happened to have in his pocket, and before I could say a word, the burst padlock was falling to the ground. I looked at him in shock and began to wonder if he really had permission to borrow this man's boat. Duncan's face contorted into a lippy smile as he winked at me. He opened his hand and showed me the man's padlock with a key sticking out of it.

'If we get caught,' he said nodding at the burst padlock lying at his feet,' the bailiffs will think we pinched the man's boat, and no take it from him.' He stepped into the shed and quickly found the outboard motor and a spare can of petrol. He handed me out the oars and rowlocks. I went down the beach and fitted them in the boat. Within ten minutes, we were ready. Duncan pushed the boat out, swung his backside onto the bows and then lifted his feet aboard as we silently slid out into the black water. As we drifted away from the shore, I saw water seeping in between several of the old planks. I put my hand on the old rusty paint tin that served as a bailer, and slipped it under my seat in readiness. I lifted my eyes and stared out into the oily blackness of the loch as the sounds of the night filled my ears.

Duncan, presumably hearing no man-made noise stirred, he nimbly stepped over the thwarts and seated himself in the stern. A moment later, the outboard spluttered into life and we were heading out into the open loch. On the far north shore, we could see the odd car headlights flickering through the trees as they wound their way along the twisty road.

Loch Ness was about two miles wide at this point, and I was, as you can imagine, extremely nervous. I had heard stories, lots of scary stories, about the Monster, and as I nervously scanned the dark waters, I imagined every ripple was the, 'Beastie,' just below the surface. Duncan though, seemed quite relaxed. He sat puffing away at his unlit, and by now soggy cigarette. As we approached the far shore, Duncan slowed the outboard to a tick-over. A few minutes later, he shut it off, changed seats with me, and rowed the last hundred yards or so into the bay.

As we passed under the crumbling battlements of Urquhart Castle, my state of unease increased, and when an owl hooted close to my ear, I almost jumped out of the boat in fright. I looked up and saw the silent ghostly white bird, as it disappeared into the loch side wood. Duncan stopped rowing, and as the small boat drifted on, he delved into his pocket, removed a tobacco tin and stowed away his soggy cigarette. When he was satisfied we were alone, he rowed the boat in close to the shore.

'Now,' he whispered, and as Duncan gently pulled on the oars, I fed the net into the still water and watched as the cork floats streamed out over the black surface, then disappeared into the gloom. He rowed the boat in a circle, judged it just about right, for as the last of the net ran out he managed to pick up the beginning. Duncan quietly shipped the oars and we drew the net closed. Then we started to haul it in, me on the sole rope and him on the float line. But the net was light and empty, no fish atall, not a solitary one.

'Thought you said the bay was full of fish,' I whispered.

'The monster must have eaten them all,' he whispered back. His attempt at humour was lost on me; for I thought he could be right.

'We'll try again over yonder,' he whispered as he quietly rowed the boat across the bay, then he backed in close to the shore. 'Now,' he said as we felt the boat touch bottom. I started to feed the net out again as he rowed in a big U back to the shore. When we retrieved the net, we removed three fine silver salmon.

'Once more, a wee bit further out,' he whispered, so we tried a circle again. But when we began to retrieve the net, it became very heavy. I could feel the ensnared fish struggling deep down in the black peaty water. Suddenly we heard a loud crash on the shore and froze with our guilty hands on the illegal net. We stared at the shore and waited for the shout or the spotlight; but none came. The bark of a roe deer a second or two later almost stopped my heart. I'm getting too old for this, I thought as we started to retrieve the net again. Then all of a sudden, the net gave a mighty tug and became incredibly heavy.

'We must have twenty salmon or maybe more,' I whispered. Then the net stopped and I felt several heavy tugs that almost ripped the net from my hands. Uninvited, a vision of the monster entered my head and sweat formed on my brow. I heaved and the net became light. It came in fast. My fear increased as I pictured the Beastie racing up towards us to take her revenge. Then right next to the small boat a big black shape leaped out of the cold water. It sent an icy spray of peaty wetness over me. In a state of total panic, I dropped the net. Dived to the wet floor, and put my hands over my head. A moment later, a wave broke over the side and sloshed aboard.

With my eyes screwed tight shut I felt the small boat rocking and pitching. I lay shaking with fear with the horrible visions of the monster in my head. I lay in the cold wetness for an eternity, unable to move. But gradually the small boat steadied and a deathly stillness settled over the loch. I slowly forced my eyes open and in the eerie stillness, I heard the hoot of a far away owl. Slowly I gathered my courage, lifted my head and ever so carefully peeked over the gunwale. I was half expecting to see Duncan's mangled body in the monsters jaws, with his lifeblood dripping into the water. Instead, all I saw was a small black slimy tree branch with a knobbly end. It lay on the surface all tangled up in the net. I breathed out a great sigh of relief, and feeling extremely foolish I glanced at Duncan; fully expecting one of his cutting, sarcastic comments. But Duncan sat perfectly still. As if frozen. As if dead. His eyes gaped wide and stared out into nothing. A small trickle of drool seeped from his toothless mouth and hung like an icicle from his stubbly chin. In the moonlight, I saw that his white knuckled hands were still tightly gripping the net.

'Duncan, are you all right?' I whispered moving closer, but I received no reply. I leaned over and shook him. He felt rigid. Every muscle in his body seemed clenched tight. I took hold of his ice-cold hands and forced his fingers free from the float line. I hauled in the net and eventually removed two small black salmon. After dispatching the fish, I none to carefully cut the slimy old wood free and watched as it slipped quietly back into its peaty grave. During this time, Duncan sat deathly still; he never moved his rigid body, uttered a sound or altered his fixed stare.

I bailed the cold water out of the boat, started the outboard and slowly began to motor back across the loch. Suddenly the moon slid back behind its cloud and the darkness descended on me like the shroud of death. In the eerie blackness, I thought about the big wave that washed into the boat. Then as my mind relived the recent events, I thought about the wee tree branch and the two fish. I foolishly asked myself why the net felt so heavy. Then the picture of Duncan's terrified expression filled my mind, and I remembered that he was not looking at the water, but looking up towards the sky.

The small boat suddenly rocked as if caught in the wake of a big ship. As I increased speed, my fear grew to near screaming point. I carried on into the night whistling a tuneless tune and trying to think of anything except the monster.

At long last, I reached the south shore and searched in the darkness for the boat place. And as I slowly motored along the shore, I wished I had paid more attention. The moon suddenly blinked on, and in its dim light, I saw Duncan. He sat deathly still, but now his hands rested on his lap.

I eventually found the boat place, ran the boat ashore, then dragged Duncan out and sat him against a tree. I hauled the boat up; made it secure, returned the outboard and the oars then relocked the shed. The net, I stuffed into its bag and hid in the loch side wood. The fish, I put into Duncan's van. Fortunately, Duncan himself was not a very heavy man, so I heaved him over my shoulder and carried him up the bank and sat him in the passenger seat.

The drive home was a strange one with Duncan sitting silently staring straight ahead. When I got home, I roused my wife out of her bed and got her to drive our car, while I drove Duncan's van, minus the fish, to the village phone box. I called an ambulance and we abandoned the scene.

I visit Duncan now and then, although I'm sure he has no idea who I am. He has never spoken a word since that night; he just sits in his chair, wearing a plastic bib and staring at whatever it was that deprived him of his wits. He's just a mystery to medical science.

End

A Night out, is loosely based on a true story that was told to me by an old rogue, and 'long retired', poacher.

The Loch Ness Monster or 'Nessie'.
Gaelic word, 'Niseag.' Or as some of the locals call her, 'the Beastie.' The first known published mention of the Monster was in the, 'Life of St. Columba,' written by Adamnan, around the 7th century. It describes how in the year 565, Columba, saved the life of a, 'Pict.' Adamnan described the event as follows:-

The first recorded sighting.
"(He) raised his holy hand, while all the rest, brethren as well as strangers, were stupefied with terror, and, invoking the name of God, formed the saving sign of the cross in the air, and commanded the ferocious monster, saying, 'Thou shalt go no further, nor touch the man; go back with all speed.' Then at the voice of the saint, the monster was terrified, and fled more quickly than if it had been pulled back with ropes, though it had just got so near to Lugne, as he swam, that there was not more than the length of a spear-staff between the man and the beast. Then the brethren seeing that the monster had gone back, and that their comrade Lugne returned to them in the boat safe and sound, were struck with admiration, and gave glory to God in the blessed man. And even the barbarous heathens, who were present, were forced by the

greatness of this miracle, which they had seen, to magnify the God of the Christians."

Pictish carving.

Stone Monster Some people say this 1500-year-old carving is the first picture of the Loch Ness monster. Hmmm, I myself think it's more like a Springer Spaniel. What do you think?

Urquhart Castle.

Our poachers set their net under the Battlements of Urquhart Castle. The present Castle, 'built for the son of King Alexander of Scotland,' dates from around 1230, and destroyed in 1692

Highland cattle once roamed the hills and glens of the Scottish Highlands. At **Lammas time** (the 1st of August.) the cattle were herded together, sorted out, and the unlucky ones driven south to the markets. The word appears in the ancient Celtic calendar. It refers to the midway point between the Solstice and the Equinox.)

The drover trails meandered their way through the glens and high passes towards the coast. One of the trails ran along the south shore of Loch Ness, passed the village of Dores, and on to the market at Dingwall, Inverness or Aberdeen.

Drover stone.

A reminder from the past is a partly submerged stone that can be seen at the west end of the village.' This stone was an important indicator for the drovers heading for Dingwall. When the stone was submerged, the River Ness was too deep to cross.

Tradition.

The wives of the returning cattlemen, (who spent the summer in the high sheilings, or bothies, minding their cattle,) made Crowdie. Although this tradition is all but gone. You, yes you, can easily make this lovely soft cheese, and here's how.

Crowdie

You will need about two pints of full cream milk, a wee bit of cream and a pinch or two of salt. If you are in a hurry, you can sour milk by adding a squeeze of lemon juice, but this might taint the Crowdie. It's best just to leave the milk in a warm place for a few hours. When you have freshly soured milk, pour it into a pan and gently warm. (Do not boil, or simmer.) When the milk curdles, (separates,) remove it from the heat and allow it to stand until cool, but not in the fridge.

The next thing you need is a piece of cheesecloth. (OK you can use something else, how about an old threadbare pillowcase.) My mother placed her cheesecloth into an old green colander, (but I don't think the colour's important,) then she tipped the curds and whey into it and left it to drain.

When the liquid has drained away, gently squeeze out the last of the whey by hand. The stuff you want is in the cloth.

Almost there, now place the crowdie into a bowl and mix in a little salt and a little cream. Finished; now pop it into the fridge to cool. Told you it was easy. You can if you like, omit the cream, or pack the cheese into a mould. It also freezes well.

The crowdie has a light delicate flavour that I'm sure you will enjoy, and what fun it would be, to say nonchalantly to your guest. 'Oh do try my cheese, I made it this morning'. The best way to eat Crowdie is on oatcakes; Treat it like a mild soft cheese.

Now, as you went to the bother of making your Crowdie, wouldn't it be nice to serve it on your homemade oatcakes or eat them with soup, stovies, crowdie, honey etc. To make oatmeal, put rolled oats, (Porridge oats) into your blender and pulse a few times. (Works fine.) The more you pulse the finer the oatmeal.

Oatcakes
1 cup of oatmeal.
1 cup of plain flour.
½ cup of milk.
Tablespoon of soft brown sugar.
75g butter or margarine.
1 Level teaspoon salt. (You can use less.)
Level teaspoon of bicarbonate of soda. (Baking soda)

Sieve the dry ingredients into a bowl, add the oatmeal, and mix. Cut the butter (or margarine) into small pieces and rub into the mixture. (Yes with your hands.) Add the sugar and mix in, add the milk until you have a stiff workable dough. It might be a bit sticky for a few minutes so do not rush to add flour. Flour your baking board, turn out the dough and sprinkle some flour over the top. Roll it out to about ¼" inch thick. My mum used to prick all over with a fork. Cut out with a cup or cutter, into biscuits, place on a greased baking tray (just lightly rub over with the butter wrapper). Pop into a pre-heated oven '180°c' and bake for 15 to 20 minutes. Lift the hot biscuits onto a wire-cooling tray with a fish slice. Store in an airtight container.

As made by the wives of the drovers a thousand years ago.

Ancient oatcakes
200g medium oatmeal.½-teaspoon salt.
1 tablespoon melted lard, or butter.
Cold water from the well.

Mix the dry ingredients together by hand. Work in the butter with your fingers. Add, 'water from your well' a little at a time Do not make wet. Sprinkle some oatmeal onto your board, and roll mixture into an almost dry ball. Press down to form a round, about a ¼" thick. Cut into wedges and cook for five minutes, on a hot girdle over the peat fire. Now all you have to do is dry them in front of a fire for about a quarter of an hour.

Cooking time will vary depending on heat and thickness. Don't worry about over cooking them, and turn them over if you like. A good, authentic taste but not as soft and light like the first recipe, I like them for a change now and then.

<u>In The Village Inn</u>.

It had just gone six o'clock on a warm summers evening, when half a dozen strangers noisily wandered into the village Inn. Jock lifted his head and slowly scrutinised the newcomers and recognised them for what they were. Tourists, American tourists.

The visitors were at first a little disappointed to find only one customer sitting by himself nursing half a pint of beer.

'Not exactly a hive of activity,' one of the newcomers said in a stage whisper. Ella the barmaid laid down her newspaper and welcomed the early customers. While she was pouring out an assortment of refreshments, one of the visitors, a tall man with rather a round face and wearing a flamboyant yellow shirt, asked her if she was local.

'Oh no,' Ella replied, 'I'm an incomer; my parents moved here from Glasgow in the nineteen forties.'

The visitor thought the barmaid was pulling his leg but he decided to ignore her comment and ask his question anyway.

'An ancestor of mine,' he began, 'came from here abouts. His name was Shaw. Are there any people of that name still living in the village?'

'You could try asking Jock over there,' Ella replied nodding towards Jock, 'his family have lived here forever.' The visitor glanced over towards the man sitting by himself by the unlit fire.

'Thank you Ma'am I'll do just that,' the visitor said, 'in fact I'll go over and ask him now.' He turned and wandered over to Jocks table. 'May I buy you a drink sir?' the visitor with the yellow shirt asked the man who was now sitting looking into his empty glass.

'Well now, that's very good of you,' Jock replied as he raised his voice a little and spoke directly to the barmaid. 'I'll just have my usual Ella, and a wee chaser,' he added glancing sideways at the visitor. 'Now why don't you all come over and sit here by me,' Jock continued, the visitor stopped wondering what a chaser was and pulled out a chair. 'It was Shaw you were asking about, was it not?' Jock said, eyeing the visitor, 'now would that be the lowland Shaw or the Highland Shaw? Well as you're here looking for your people you must think your ancestor came from hereabouts. Now that would make you a Sithech Shaw. Sithech means wolf in the Gaelic.

The visitor's eyes sparkled with excitement.

'Aye your people more than likely came from Tordarroch, that's just over the hill yonder. They're also known as Clan Ay, or children of Shaw, and your clan motto is "Fide et fortudine" that's by faith and fortitude, in the English.'

'Yes, the American replied with enthusiasm,' my ancestors name was Thomas, and he was born here in Dores in 1798.'

Jock lifted his head and closely scrutinised the visitor's features. The man in the yellow shirt felt a little uncomfortable, so he glanced towards his fellow travellers for moral support.

'He was christened at Dores in November 1798. I have a copy of his birth certificate,' he said as he leaned back in his chair.

Jock eventually released the man from his blue-eyed stare and glanced over towards Ella, and watched as she poured out a double dram of Genfiddich. 'We had a lot of families of that name around here at one time,' he continued slowly without taking his eyes off the glass of amber nectar. Then when she placed it on the bar counter, his face distorted into, what the watchers took to be a grin and he began to rub his chin and lick his lips. The man in the yellow shirt rose from his chair and went over to the bar to pay for the drinks. When he returned, he placed a pint of beer and the glass of whisky on the table in front of Jock.

'Thank you very much,' Jock said as he carefully lifted his dram. 'Your very good health,' he added as he downed his whisky in two slow swallows.

As the visitors sat down around the table and introduced themselves, Jock shook the last drip of Genfiddich into his beer. They told Jock in voices that would have carried to the far corners of the village hall that they came over from America on holiday to visit the old country, and to look for their roots. The man from America with the yellow shirt repeated his question.

'You're a Shaw then,' Jock said shaking his head. Ella glanced over and exchanged looks with Jock. 'It's quite a popular name in the Highlands you know. We did have a few families in the village and along the Loch side by that name.' 'A hermit lived in a wee hut above the loch,' he continued as he leaned back in his chair. 'And an elderly couple lived in a wee cottage at the west end of the village when I was a boy. But they and their cottage are long gone.'

'What about the other families?' the spokesman asked with perhaps a little less enthusiasm.

'At one time,' Jock continued in his quiet Highland voice that made the visitors lean forward to listen, 'crofts occupied the lands along the shores of Loch Ness. The crofts are nearly all gone now, swallowed up by the forests that now cover the hills.

'On one of the pre' forest crofts, there lived a spinster woman and her bachelor brother. Now they went by the name of Shaw. I'm likely the last person that remembers them. She had a wee catch phrase. Fine and handy she used to say, fine and handy. They scraped a living out of the steep stony fields. They had a few sheep, a puckle (a few) hens and it seems to me that they may have had a milk cow; but I can't be certain.'

'My late father told me, that when he was a boy he would see her riding her pushbike, going, or returning from Inverness with her messages, (shopping bags) hanging from her handlebars. The thirty mile round trip was a fair pedal. At that time, the road was narrow and twisted, with grass growing down the centre, and a good few braes (hills) on the way. Teams of horses plied this narrow road pulling drays loaded with timber. This was well before my time of course,' Jock pointed out as he took a drink from his beer. 'They must have been a constant nuisance forcing her to dismount to let them pass. Did she complain, not her. Och, she would say, the bicycle is fine and handy. The annual ploughing was always a bit of a chore for them, she would put on the harness and drag the plough back and fore, while her brother worked the long plough handles. It must have been backbreaking work for them both. Did she complain? Not her. Och the plough's fine and handy.

Her brother, who was a few years older than herself, had a wee shed outside the thatched croft house. Now in this shed was a stick-horse; you know the thing, you would lay a log on it so you could cut it up with your saw for firewood. Well the stick-horse had another use, for a Brown Bess musket lived on it. It lay loaded pointing down towards the field. The musket had been in their family since the battle of Culloden. Mind you, my grandfather had modified it a bit for him. My grandfather was a water bailiff and lived in the next croft along. He, my grandfather that is, removed the old flint lock, drilled out the touch hole, tapped the now bigger hole and screwed in one of the new cap-locks. A big improvement, the crofter could now leave the gun loaded and ready to fire. The wee copper 'top-hat,' detonator was impervious to the damp. So I imagine venison was a regular part of their diet.'

'The lady herself had a wee hobby too. On the fine nights, she would sit by the door of the croft house spinning away on her mother's old spinning wheel. Rocking away with her foot as she drew out a fine strand of woollen thread. She would lift her head now and then, to take in the view. She could see across, and down Loch Ness in both directions. She would more than likely have sat by her door in her last years and watched John Cobb, as he raced his powerful speedboat up and down the Loch as he tried to gain the world water-speed record.'

'She had a sort of loom, as I recall. It was a bit worm eaten and was held together with bits of twine. She would weave their cloth and make some of their own clothing. Man, the sister was a dab hand at the sewing. I remember calling by as a young boy with my father.' The storyteller paused and rubbed his stubble beard as if trying to recall the occasion. 'I think it must have been a special day,' he continued a few moments later, 'because they were kinda dressed up in their best clothing. As I look back into my memory, I can see them as very old. Perhaps in their eighties; but then I suppose small boys tend to think everyone taller than themselves are old, so they may well have been a little younger; but I don't think so. In fact they may have been even older.' Jock added nodding his head. 'You know I'm thinking it might have been the Coronation. Yes that's what it would have been, Queen Elizabeth the 6th Coronation.' The deep meaning of Jocks statement went over the visitor's heads, but Ella grinned.

'The spinster, oh yes I was telling you about her sewing. The day we called, the brother wore a nice pair of trousers, with just the tiniest wee flaw. They were made of two older pairs and sort of, joined together in the middle. If they had not been different colours, no one would have noticed. They invited us in for a cup of tea. We sat down on the wooden chairs by the fire. I clearly remember looking at the clootie rug that lay on the floor.'

'What's that?' one of the visitors asked.

'Oh, a clootie rug, well I'm sure you got the rug bit, so it must be the clootie you're wondering about. Well a cloot is just a bit of old cloth, tear it into strips, then knot them through a sack to make a rug, a clootie rug. An old Highland saying is, 'never cast a cloot until May is oot.' This is of course a reference to winter woollies, or winter clothing, Jock explained. 'Or a clootie tree,' Jock said very much enjoying the yarn. 'If you had an illness or sickness you would tear a cloot from your clothing and tie it on to the branch of a special tree, as the cloot rotted the illness went away. Many people still swear by this cure. How about a clootie dumpling, aye, you've got it now, a dumpling tied up in a cloot and boiled.'

'Och, sorry,' Jock said, as he picked up his empty whisky glass, looked at it then laid it down again, 'I got a wee bit side tracked. I was telling you about my visit to the Shaw's croft. We were, as I said sitting at the table, it was set with cups and saucers and plates all laid out on a neat white tablecloth that was embroidered with bunches of bluebells. The effect was lost to a certain extent by the tin milk pail that also sat on the table. She had her griddle on the fire; it was hanging from a chain that dangled down from the chimney, and on a trivet sat a big black cast iron kettle with steam gently drifting out of its spout. She made the tea by tilting the black kettle with a poker and pouring the boiling water into a brown enamel teapot with a chip on its spout. I sat quietly, under strict orders from my father to behave, or else. A warning I dared not ignore. My father, only just tolerated children, but that is another story altogether. I sat looking at the old newspapers that covered the walls. A bit like higgledy-piggledy scraps of wallpaper, mostly the Inverness Courier. I remember they dated from the eighteen hundreds. She handed me a hot scone and pointed to a jar of her wild blaeberry jam that sat before me.'

'Blaeberries?' one of the visitors repeated in way of a question.

'You don't know what blaeberries are? Och, they grow on heather size bushes on the loch side and up on the hill, they have wee green leaves and black berries, about half the size of peas. We used to pick them in the autumn. I remember they were bad for staining my hands and tongue; turning them purple. I think you might know them as bilberry's, or is that another fruit all together. Och you'll no doubt know best yourself,' he added as he picked up his empty glass and looked at it. He laid it down with a faint sigh, then drew out his pocket watch and scrutinised the dial.

'Will you have another drink?' the visitor in the yellow shirt asked.' Thinking that Jock was about to leave.

'Och well, I'll no insult you by refusing, I'll just have the usual,' he said aloud to Ella. 'Talking about fruit,' Jock continued, 'reminds me about the wild Arctic strawberries that grew on the gravel beach by the loch side, tiny wee red berries with an unforgettable taste.'

The visitor returned from the bar with Jock's refreshments.

'Ah thank you,' Jock said as he downed his dram, then once again shook the last drip into his pint.

'Anyway as I was saying,' he continued after wiping his mouth with the back of his hand. 'The brother sat blathering away to my father for a time. Then he carefully picked up his cup and poured his tea into his saucer. Then using both hands he raised it to his lips and noisily slurped his tea.' Jock demonstrated by lifting his pint in both hands and noisily slurping his beer. Ella looked over at the enthralled audience and grinned. 'I learned later,' Jock continued in his soft Highland voice, 'that the tea cup was not normally used by them, and they only brought them out on special occasions. They usually drank their tea from bowls.'

'After tea, they took us on a tour of the croft,' Jock stopped talking and rubbed his stubbly chin. 'I remember now,' he said thoughtfully. 'They did have a milk cow; it was standing looking over a fence at us. The sister said it wanted milking. It was the brother himself, who showed us the wavy drills of tatties. The shaws (leaver or stalks of potatoes etc.) were starting to wilt, and I saw that one of the drills was just about used up. A grape stood by the next plant in line, and a white chipped enamel pail sat next to it. The brother dug the plant up and shook the tatties free and we filled the bucket with the multi sized earthy tubers; he then poured them into a sack and handed it to my father; a wee gift for us to take home.' Jock stopped talking and shook his head, 'no it was not the Coronation. That took place in June I think.' He looked up at the tourists puzzled faces. 'The tatties! We dig them up in late summer and autumn. Anyway the reason I remember that visit so clearly,' Jock said 'is something the old lady said, she got quite excited when she told us.'

'Do you know?' she said waving her hands in excitement. 'There's a wee shop opened up in Torness. Just over the hill behind us,' Jock explained, pointing over his shoulder. 'It'll be right fine and handy.'

'When I drive down the loch side.' Jock said slowly,' I always look up at that forested hillside that covers their croft. I think of her climbing over the steep mountain to the glen at the other side. A trek of at least four hours if not half a day.

'But I don't think this was the Shaws you were looking for, they were the end of a line, and when they went to their maker, the croft was never re-let.'

'Were there a lot of crofts on the loch side,' a visitor asked.

'I seem to remember,' Jock replied, 'three crofts on that particular part of the hillside; the next along at that time was tenanted by a friend of my fathers. A man by the name of, Lachie Stuart. My grandmother, the previous tenant had long since moved along the loch side to the village here.'

'When was that?' a curious voice asked.

'Och they moved about nineteen sixteen, my grandfather died young and my grandmother moved into the village and later married the butler from the castle.'

'My father, as I was saying, and his friend Lachie, would go out on the hill shooting now and then or go fishing in the loch for salmon. I think I only met this man the once, when he came into my mother's post office. Come away through she called, come and meet this man; this is the man that took the photo of the monster. The one with three humps that is. The story, as told to me was that he was rather a canny man. He saw no reason for his milk cow to be eating his grass when there was perfectly good feeding for her on the loch side. The only trouble was he had to go and find her at milking time. One morning he went out. Or so the story goes. To fetch her, and saw the monster out in the loch. He ran home for his, 'Box Brownie' camera and took a couple of snaps. He then came along to the village and told my father, who took him into Inverness on his motorbike and they sold the camera; complete with undeveloped pictures, to the local newspaper. I believe they got fifty pounds for it, not bad in nineteen fifty three. Later, when Lachie was an old man, he confessed that his photo was a fake. Just a few straw bales covered in black canvas. I have no idea if my father was part of the scam. I suspect not, for he was an honest man. Sadly, he is no longer here to ask.' Jock paused for a moment and looked around the bar, then his eyes came to rest on Ella and he grinned.

'Speaking of the monster,' he continued 'reminds me of another man. Dodd, his name was.' Ella glanced up at the mention of her late father, 'he came running along to our house one Sabbath,' Jock raised his voice slightly for Ella's benefit. 'He was very excited and shouted that the monster was out in the bay so we all ran out the back door to have a look. And sure enough, there was something out there, but it was a bit too far away to see clearly. My father went back into the house for his telescope and spied the, 'beastie.' Och, he said it's only a couple of deer swimming across the loch, here have a look,' he said offering the man the glass. (Telescope) 'Not at all, the man said, that's the Monster, and he turned on his heels and walked away.'

'Since I mentioned the Loch Ness Monster I suppose you want to know if there really is a Monster living in the Lochs peaty depths. Well the honest answer is that I just don't know. I am however convinced that there is something strange out there.'

'Let me tell you what I know,' Jock said lifting his finger and waving it in the air, 'there are enough eye witness reports to convince any court of law, and not just from tourists like yourself.'

'My father, twice in his lifetime hooked something when fishing that was bigger than any salmon. So big in fact, that the line stripped from his reel in a few seconds, ran out then snapped. He said he was lucky the rod never broke. I know for a fact that there are some very big eels out there, and some times a seal finds its way into the loch. Who knows what else, but as far as photographic evidence is concerned, I'm a bit sceptical. For as we all know the camera is a dreadful liar. Perhaps the biggest thing against there being a monster, of the prehistoric type, is time. Fifteen thousand years ago, Loch Ness had a mile of ice and snow piled on her, so the monster would have had to come from the sea. Well why not, all things are possible.'

By now, the village Inn was filling up and Jock was constantly nodding a greeting to customers as they entered. One of the visitors placed another round of drinks on the table and Jock picked up his dram.

'I'll tell you about a man who worked with my grandfather; he was also a water bailiff. One of his jobs was to row a fourteen-foot rowboat west up Loch Ness, that's against the prevailing wind. No outboards in those days, so the Laird or his guests could fish down with the weather. Well one day this man was rowing up the Loch when the Monster rose beside the wee boat and gave him a terrible fright, he put the boat ashore and refused to go on the loch again. That's a true story told to me years ago by my old aunt who was in the house when he came in. White as a sheet he was she said, and shaking all over, it apparently took two good drams to bring him round, but he never did go out on the loch again.'

'There are lots of stories like that, and a lot more recent ones too. I met a man once who dived in the loch, SCUBA I mean, he told me it was a weird experience, not only was it very cold and dark but the visibility was further impaired by suspended peat particles in the water, that made his torch useless. He said he would not be doing that again.'

'You'll no doubt be taking a drive around the loch,' Jock said looking at the visitors. 'This south shore is the quiet side, with lots of places to stop. Get out of your car,' he said quietly, 'and sit on the beach on your own. Look up at the high mountains and down to the dark water. Remember the long gone people like the brother and sister. Picture the drovers, with their plaidies wrapped around, who for thousands of years drove their red Highland cattle along the loch side. Think about the Highland clans marching to battle with their long claymore swords. Relax by the Loch side, don't rush, this is a magic place. Let the past come to you. Oh and don't forget to have your camera at the ready, you never know it might just be you, that gets that perfect picture of, Nessie.'

The visitors eventually left the Inn and Jock returned his empty glass to the bar.

'Well, Jock Shaw,' Ella enquired. 'Were they relations of yours?'

'I'm sorry to say that the one with the yellow shirt is a sort of cousin of mine,' Jock replied. 'His great grandfather left the village in disgrace.' Jock leaned forward and whispered in a voice that only Ella could hear. 'He married a Campbell you know.

End

The yarn Jock spun in the Village Inn was about real people who lived and worked on the loch side.

Just in case, you're wondering about the slight discrepancy regarding the title of the British and Commonwealth Queen. She is of course Queen Elizabeth the 6th of Scotland, and the 2nd of England.

VIEW OF LOCHNESS MONSTER, TAKEN NEAR DORES

Lachie Stuarts 'Straw Monster photo'.' For many years, it was on display in the old Dores Post Office.

John Cobb.

Who once held the land speed record, raced his jet-powered speedboat Crusader on Loch Ness, where he reached a world record speed, in excess of 200 mph. In 1953, he died when his boat hit a wake and broke up. His boat is still down in the lochs peaty depths.

Some of the measured mile marks are still standing and there is a monument in his memory on the North shore.

Murder of Glen Coe.

Jocks whispered statement, 'He married a Campbell you know,' stands on its own and to most Highland Scots; it will require no further explanation. He was of course referring to 'Mort Ghlinne Comhann.'

In the winter of 1692, a band of men under the command of Robert Campbell, marched into the glen of Glencoe. They asked for hospitality from the houses of MacDonald at Invercoe, Inverrigan and Achnacon. For thirteen days, they lived among their hosts as honoured guests, then at five o'clock one morning they rose and murdered 38 members of the MacDonald clan. 40 more people, mostly women and children died in the snow after fleeing.

This crime was so devious and treacherous that even today some Highlanders will shun the company of a Campbell. Or ostracise them if they dare marry one. The reason given for the massacre was that, the MacDonald clan was a bit slow in pledging their allegiance to the new joint monarchs, King William the 3rd and Queen Mary the 2nd.

The background to this massacre. Involved the monarchs, politicians, bribery and corruption. As usual, religion played a leading role. Ironically, very few Campbell's took part in this crime; most of the men involved were soldiers of the crown.

This event have been immortalised in this song' By (Jim McLean.).

The Ballad of Glencoe

(D) Oh, cruel was the snow that (G) sweeps (D) Glencoe.
And covers the grave o' (A7) Donald
Oh, cruel was (D) the foe that (G) raped (D) Glencoe.
And murdered the (G) house (A) of Mac (D) Donald.

They came in a blizzard, we offered them heat.
A roof for their heads, dry shoes for their feet.
We wined them and dined them, they ate of our meat.
And they slept in the house of MacDonald.

O, cruel was the snow that sweeps Glencoe.
And covers the grave o' Donald.
O, cruel was the foe that raped Glencoe.
And murdered the house of MacDonald.

They came from Fort William with murder in mind.
The Campbell had orders King William had signed.
"Put all to the sword" these words underlined.
"And leave none alive called MacDonald."

O, cruel was the snow that sweeps Glencoe.
And covers the grave o' Donald.
O, cruel was the foe that raped Glencoe.
And murdered the house of MacDonald.

They came in the night when the men were asleep.

This band of Argyles, through snow soft and deep.
Like murdering foxes amongst helpless sheep.
They slaughtered the house of MacDonald.

O, cruel was the snow that sweeps Glencoe.
And covers the grave o' Donald.
O, cruel was the foe that raped Glencoe.
And murdered the house of MacDonald.

Some died in their beds at the hand of the foe.
Some fled in the night and were lost in the snow.
Some lived to accuse him who struck the first blow.
But gone was the house of MacDonald.

O, cruel was the snow that sweeps Glencoe.
And covers the grave o' Donald.
O, cruel was the foe that raped Glencoe.
And murdered the house of MacDonald.

Vitrified fort.
The remains of a Vitrified fort that is over 3000 years old, sits on a hilltop above Loch Ness. The prehistoric people who built this fort were obviously well advanced and organised. It is still unclear how this ancient people achieved the heat required the vitrify their fort walls, but the Highlands were covered by the Caledonian forest at the time so perhaps it's not a mystery after all.

From the Oxford dictionary, (Vitrified, 'having the form or appearance of glass.')

Evidence exists that people have lived in Scotland for over 8000 years. Farming took root, around 6500 years ago. The farmers grew wheat, barley, rye, or reared cattle and sheep.

Jock was a bit confused about Blaeberry, so just to avoid confusion. Scottish blaeberries (vaccinium myrtillus) are smaller than blueberry, and can sometimes be a little on the 'tart,' side.

Blaeberry jam.
1 kg blaeberries,
1kg sugar
1 or 2 stalks of rhubarb, or about one quarter of the total fruit.

Chop up the rhubarb and cook gently until soft. Add the sugar, slowly stirring with a spirtle, or wooden spoon. When all the sugar has dissolved, add the blaeberries. Bring to a rolling boil. Test after fifteen minutes by dripping a little of the jam from the spirtle onto a cold saucer. If the jam is still runny, cook for another five minutes and try again.

My favourite use of this fruit is in Blaeberry crumble. Add a cooking apple for bulk, but I am sure I don't need to tell you how to make crumble.

The Half way Inn.
Not far from the Shaws croft stood a coach stop and a small Inn. A place where travellers could change their hired horses. A small stone marks the site; it commemorates a visit by James Boswal and Dr. Samual Johnson in the seventeen eighties.

There's a wee Legend associated with this loch side Inn.

Robbery at the Half way Inn, or Change house.
Thieves robbed the Inn one night, and then fled west along the loch side. By the time the robbers reached Fort Augustas, they had hidden their booty. Apparently, a coachman recognised the robbers; they were promptly arrested, and hung. The booty, as far as I know, awaits discovery. My grandfather, my father and yes me, have all searched for the treasure, no luck so far but I bet it's still there.

Jock mentioned the Clootie dumpling, here's how to make it.

Clootie Dumpling
500g self raising flour.
125g fresh bread crumbs.
125g sugar.
125g shredded suet.
½ teaspoon salt.
1 teaspoon mixed spice.
1 apple grated.
250g currents.
375g raisins.
½ pint milk.
1 tablespoon treacle.

Mix flour, breadcrumbs, sugar, shredded suet, salt, and mixed spice. Add grated apple, currants and raisins. Mix well. Stir in milk and treacle until well blended. Scald coot (cloth) in boiling water, dust with flour. Place mixture on the cloot. Tie up securely, but leave room for it to swell.

Place a plate in the bottom of the pan; carefully cover with boiling water to about ¼ of the way up the pan. Lower the pudding onto the water, onto the plate. Top up the pan with boiling water; avoid poring over water over the pudding. Boil for about three and a half hours. Top up with boiling water as required. When Dumpling is ready, remove cloot. If dumpling is wet, dry it in the oven.

Serve with custard. Leftovers. Slice and fry in butter, then sprinkled with caster sugar.

The Grouse and Claret.

One afternoon in mid August, I stood at the back door of my cottage looking out over Loch Ness. For a while, I watched the gentle waves as they broke and washed up on the gravel beach. They were making that most restful of sounds that had been with me since my childhood. I heard a plaintive cry and looked up to see a pair of black and white oystercatchers swoop down from the clear blue sky and perch on the end of the old wooden pier. I glanced across the loch at the far mountains and saw that over the last few days, nature had painted the hills purple; the ling heather had silently crept into bloom.

I have always stayed by the loch side, in this very cottage, as had my father and grandfather before me. I had just returned from Inverness, the town was busy and the sheep market bustled with farmers from all over the Highlands. Some were buying some were selling, while others just came to lean on the rail and gossip with their cronies. I was there to sell my five pet lambs. I had bought them for pennies on a cold winter morning from a tired and overworked shepherd. He was happy to get rid of them. He was at the height of the lambing and had little or no free time to wean hungry orphans.

My mind drifted back to this mornings events and I saw again the fat, well fed pets. They stood confused and bewildered in the sale ring. The ringmaster had prodded them with his stick to make them walk around a bit, but they just stood still and looked at him, a puzzled look on their faces.

Lot number 104 the auctioneer had called out.

The pets all seemed to turn and stare at me, their heads leaning to one side as if pleading with me to take them home. I turned away and closed my mind. The lambs had been hand reared and spoiled rotten by my childless wife, who as usual, had given them all names. Fleecy, Snowy, Sleepy, Jumper and Spot.

See they go to a good home, she said with a tear in her eye as I loaded them into my van. We reared a few every year, just a little extra income; besides, they kept the grass tidy saving me the bother. I had not the heart to eat my own sheep. In the autumn, after the 'gathering,' I would buy strangers from the shepherd, to fill the deepfreeze. I preferred mutton as apposed to the lamb. An old hill ewe is full of flavour and ideal for 'stovies' or stews on a cold winter night, and besides they cost almost nothing.

After the market, I wandered the streets of Inverness. The envelope containing my traitors wages heavy in my inside pocket. I slowly made my way to Union Street, to the fishing tackle shop, and then spent some time looking at the tray of trout flies I had come to see.

A new gamekeeper had recently moved into the area and his sideline was fly tying. His reputation had come before him. His flies were said to be the very best. And indeed, they seemed just about right to me. I spent a good half hour making up my mind, not easy with over a hundred flies in each tray. Eventually I left the shop and wandered home with six new flies in my pocket. The sold pets forgotten for a moment.

'Well?' My wife demanded as I stepped into the kitchen.

'Well what?' I asked in a questioning voice.

'My lambs,' she yelled, 'who bought them?'

'Oh yes,' I replied remembering my foul deed. 'A play farmer bought them to decorate his paddock.' I told the same wee white lie I tell her every year; it seemed to satisfy her.

I handed her the envelope containing the money and she dropped it onto the table, and I saw her eyes moisten. I quickly escaped outside and carefully unpacked my purchases. I hooked them one at a time onto the cross board on my wooden shed door. I then sorted them out into two sets of three. I was trying to decide on the perfect combination, biggest on the point and slightly smaller flies for the droppers. I always fished for trout with three flies on my cast, as had my father before me. It worked for him and it works for me.

Dan appeared silently by my side. 'What's this Eoin' he asked, making me jump, 'new flies?'

'Oh,' I said, 'yon new keeper tied them.'

Dan leaned forward holding his thin bony hands behind his back, tilted his head and peered as if down his nose at my new flies, then scrutinised each one in turn. He spent slightly longer looking at the 'grouse and claret'

'Och Eoin man, they are fine flies, fine flies indeed,' he said nodding his head.

'And what did the man charge you for them?'

'Och they were not very dear at-all,' I replied, refusing to satisfy his curiosity.

Alex, Dan and myself went fishing on the loch at least once a week, whenever the weather was just right; and today it was about perfect. The trout would be in fine form. We usually fished in the long summer evenings, sometimes not getting home until after midnight.

My wife came out the back door with two mugs of tea and a plate of freshly baked scones; dripping with heather honey from my own hives. We sat on the old log by the beach, enjoying the quietness. As we gazed out over the dark loch, I told Dan I had sold my sheep.

'Oh?' he said as he retrieved his pipe from his pocket, 'you would have got a fine price for them no doubt. They were in top condition,' he added a minute later, as he flicked his spent match into the water. 'I'd have bought then myself you know, but I have no deep freeze just now.'

Dan never had a freezer or any money if it comes to that. Any cash he got his hands on went over the bar in the Inn. I gathered up the mugs and the sticky plate, leaving Dan puffing away on his pipe. When I returned to my enjoyable task of making up my casts, I found Dan standing by the shed door examining my flies again. I looked and was dismayed to find one of my new flies, the grouse and claret, missing. I looked at Dan suspecting his well-known light-fingers had been at work. He saw me looking.

'Och Eoin there was a wee puff of wind while you were inside,' he commented, 'perhaps it just blew away.'

I half-heartedly searched the ground under foot realising it was a pointless task. Eventually Dan wandered away saying he had a few things to do.

'See you this evening Eoin,' he murmured over his shoulder, as he wandered away towards the Inn.

That evening when Alex arrived, we pulled the larch rowboat out of its shed and dragged it to the waterside. With all our gear loaded, we sat down on the beach to wait.

Dan always seemed to arrive, after the work was done. Alex had more than once suggested that Dan's timing was no accident. As we sat waiting, I told Alex about my new flies and told him about the missing 'grouse and claret'.

'Well Eoin I reckon I know where it went,' he said with a grin. 'Bet you a pound he produces it tonight with a long drawn out tale describing how he just happened to find it lying on the road. Or how he met a man in the pub.'

I declined the wager and we both laughed.

'Eoin, Alex. What's the joke?' Dan said as he silently walked down the beach.

'Evening Dan' Alex said, 'Did you meet anyone interesting in the Inn today.'

Dan climbed into the boat glaring at Alex and sat down holding his old greenheart rod pointing at the evening sky.

I pushed the boat out into the loch, as Alex slipped the oars into the water. He turned the boat into the waves as I prepared to fire up the old seagull outboard. I wound the start rope around the flywheel, turned on the petrol, opened the air vent, closed the choke and then pressed the priming button on the float chamber. The fuel overflowed and dripped on to the water turning it into a rainbow of blues and purples. I pulled the start rope and the outboard fired at once, sending a puff of black oily smoke drifting down-wind. I slowly opened the choke as the motor warmed up. Meanwhile Alex stowed away the oars and we turned out passed the old ruined jetty and headed west up Loch Ness. Our destination this evening was just beyond the Witches burn. Always a good starting point for the twilight drift.

When we neared our destination, I turned off the petrol and allowed the carburettor to run dry.

Silence filled the air as the echo of my now quiet engine died away. Alex slipped the oars into the water and turned the boats stern, to the shore. We sat in the troughs of the gentle waves and very slowly started our drift. Dan already had his rod out, and I heard the swish of his first cast over my head as he fished the water that was mine by rights. Within a minute, I too was fishing, forcing Dan to keep to his own water. I expected that any second the first fish of the evening would pluck my new fly from the surface and leap out of the dark water shaking its head or diving deep and pulling line from my reel.

As we quietly drifted down the shoreline, I fished from the very edge of the loch to the centre line of the boat. The odd fish that swirled to the offered flies, gave promise of good fishing ahead.

The first take was mine, a solid pull. My instinctive strike sank the hook into the hard bony jaw. The trout dived deep, then raced to the surface and leaped out shaking his head. Then it dived again, pulling line from my reel. Eventually it tired and allowed me to guide it into the landing net held by Alex. It was a fair size trout, a fine looking fish, at least a pound in weight.

The first half hour was up with only the one fish between us. I changed places with Alex, and within a few seconds, he caught his first trout. As he played it, I got the landing net ready.

'Nice fish,' I said, 'just perhaps a tad smaller than my one.'

'Not at all Eoin,' Alex argued, as he pushed his thumbs through the gills of both fish and held them up side-by-side. 'Yes I thought so; look my fish is definitely a bit deeper, so it must weigh more. What do you think Dan?' but Dan was too busy concentrating on fishing Alex water to pay him any heed. During my turn on the oars, Alex landed two more trout, both too small to keep. Dan had yet to catch his first fish and remained quiet. When Dan took the oars for his turn, he insisted in trailing his line behind the boat. Alex landed one more nice fish, and shortly after that, Dan's line snagged the bottom and we had to stop fishing as he freed it.

Later when I was in the bows fishing deeper water, the wind died to a vesper, and the quietness enveloped us. The sudden bark of a roe deer, just yards away made us all look up and we watched as he disappeared into the hazel woods followed by his white bobbing tail.

Dan wound in his line as his time on the oars neared its end. He fussed about in his old fishing bag, allowing the boat to turn parallel to the shore, spoiling our fishing. We saw Dan tie on a new fly.

'What fly's that your tying on now Dan?' Alex asked winking at me.

'Oh, just one I got today, a sort of present you might say.'

'Aye but what make is it Dan?' Alex persisted.

Dan ignored his question and continued his yarn. 'I was blathering to a man in the Inn this afternoon. A man from the south it was. Up here on holiday, he said. Well we got to talking about fishing, and you know it turned out he was a friend of yon new keeper.'

'Have you tried his flies yet? He asked me. No, I have not, I replied. Well do you know he took a fly box out of his pocket and gave me one, let me know how you get on with it, he said'.

'That's very interesting Dan,' Alex interrupted, 'and what was this mans name?'

'Ach you know I just can't quite remember,' Dan replied, 'but he was a real gentleman, yes a real Toff.'

Dan made a point of looking at his watch. 'Well, well, look at the time; it's my turn to fish again.' Alex glanced at me and I winked back.

Dan's luck changed and he hooked three trout within the next ten minutes.

'Man, that new fly of yours is working well Dan,' Alex commented, 'what did you say it was again?' The hoot of the Owl made me look up and I saw her white body silently gliding over our heads. The evening was drawing on and the hills were already slipping into darkness. We had drifted down the Loch and were just opposite, 'The Phantom Hand well'. (See below) Dan's turn on the oars had come and gone and he was again fishing the deep water. Alex was on the oars and once again, I had the best water.

Suddenly the silence and peace was shattered when Dan hooked a big fish. He stood up and shouted 'yes.' We both turned and saw his old rod bending far more than was good for it. His old brass reel was losing its line with a rapid screaming sound. Dan was trying desperately to slow the fish down before his line ran out. Alex, immediately turned the boat around, and pulled on the oars for all he was worth following Dan's fish.

'It's a really big one.' Dan shouted. 'Faster Alex' Dan screamed in near panic. 'I'm almost out of line.' But before he had the last word out of his mouth, his line ran out. There was a loud crack and his old dry rod snapped, just above his hands. The line parted a split second later. Fish, fly, line and rod all disappeared into the dark peaty water. We were all silent as we stared at the dying circle of ripples on the glassy water.

Dan sat down heavily with a sigh. Alex leaned on the oars and looked up at me and Dan looked as if he might cry.

'This was my father's rod,' he said as he examined the remaining part. 'He gave it to me the day he died.'

'Dan you're a dammed chancer,' Alex roared, turning around on his seat, 'you stole that rod from Andy Ross.'

Dan looked up, 'well maybe Andy did give it to me,' he said, 'I can't rightly remember.'

'We'll call it a night,' I suggested and Dan nodded his head. I started the Seagull and motored the last quarter mile home. We shared out our fish, three trout each. Not a bad evenings fishing. Dan wandered away and Alex and I sat down on the beach quietly talking and enjoying a wee dram from his hip flask.

When his flask was empty, Alex went home and I wandered into the kitchen and put the trout on the sink draining board. My wife came through from her TV.

'You're home early Eoin,' she said, as I started to clean the fish.

'Fancy a trout?'

No, put them in the freeze, there's some stovies in the oven for you.

'Oh' she said, 'as she opened the Aga door, you dropped one of your flies. I found it by the shed door, it's on the table.' I glanced over and saw my fly sitting on the white blood money envelope. It was my new 'grouse and claret'.

End

Legend. Well of the Phantom Hand.
Dan hooked his big fish, just below this well. The well is about a quarter of a mile west of the village of Dores. It nestles half hidden, among the moss and bracken, by the roadside.

This well has a bloody past, and more than one man lost his life while partaking of its cool waters. The murders took place in the seventeen twenties. They were soldiers under the command of General Wade. (1673-1748.) Wade was building a new road along the old drovers trail. The road, next to the densely wooded hill, was largely un-welcome. The Jacobites took exception to this intrusion into their heartland, and made their feelings felt.

That was one of the legends, my late father passed down to me while we fly-fished over the same waters.

Jacobite Graves.
The fishermen began their drift just west of the Witches Burn. In a near by gully, there are two unmarked graves. Two Jacobite Highlanders were returning home from the battle of Culloden. (1746) they died of their wounds and their friends buried them by the loch side.

Loch Ness Trout.
The best way is over an open fire by the loch side, wonderful. But you can cook it using a frying pan. First, add a large knob of butter to the pan and place it on a moderate heat. When the butter has melted, place the cleaned trout in the pan and gently fry. When you judge it is half cooked, turn it over and cook the other side. The trout is ready when you can slip a knife into the bone on the thickest part. Remove from the pan and pop it onto a warm plate, add another knob of butter to the pan and a squeeze of lemon juice.

Brown the butter and pour it over the trout. On your plate, gently run your knife down the centre line of the trout, from end to end. Then using your fork, lift and fold back the two fillets to expose the bones. Now gently lift the bone clear, with the head and tail still attached. If the meat sticks to the bone, the fish is slightly underdone. If the bones fall apart, the trout is over cooked. (Never mind you will get it just right next time.) Now fold back the fillets and remove the visible fins along with their attached bones. That's it; the only bones left are the underside fins. Now you can really enjoy your brown trout.

I have mentioned the old pier at Dores twice now in stories, so I better tell you a little about it. In the early 1900s, they cut down the forest above the Loch. They used bogies that ran on steel tracks, to transport the sawn timber, from the mill, out to the end of the pier, to the waiting boats.

The Card

It was the saddest of days. A day that lasted an eternity, a day without end, so much hurt. How could a boy bear such grief, his heart was broken beyond repair.

'There, there,' his mother said as she gave him a cuddle and tried to comfort him, 'she was very old you know. It was a kindness really.' As usual, his mother just did not understand; she had not loved as he had loved, she could not possibly understand his loss.

'Come on,' she said, 'get your school bag; you'll be late for the bus.' With that, she hustled him out the door and he slowly wandered down the lane quietly sobbing. That was eight years ago to the day, and he still missed her. His parents had gone out and bought him another dog, but it was not the same, how could it be. He had learned to show the replacement affection, and took it for long walks; but he could not love her, not really love her. He could not love any more. He had decided to keep his distance, and not expose his being to the possible grief of another loss.

Now it had all changed, he was in love again. Well not quite the same, this time it was a girl, a young lady, a beauty beyond compare. She had arrived as the first white snow of winter fluttered down from the mountains and bathed the village in a new virginal whiteness. Her father, who was the new bobby, drove into the village trailing a cloud of delicate swirling snowflakes behind his green Volvo. As the car drew to a stop, the flurry of white frozen mist drifted away in the breeze. Then he saw her. She stepped out of the car like an angel from a cloud. He was immediately captivated, smitten and imprisoned by her beauty.

She had long dark hair that drifted in the cold air as she walked. She wore makeup, and dressed like a fashion model. He was awestruck, unable to move so he just stood and stared. Her beauty knew no bounds, and he wanted to be with her forever.

In the coming weeks he dreamed of her, and not only at night but during the day as well. He could see them together, arm in arm walking, kissing and cuddling. He could even see her by his side as his wife. He wrote poetry for her, long verses of heartfelt emotion, dripping with sentiment and undying love. If only he could bring himself to speak to her, all would be well, but he lacked the courage, so he just worshiped her from afar.

Every day he saw her. Some times, she would say hi in passing, but he could never reply, his mouth would go dry and he would be unable to speak. Then as she walked away, his head would suddenly fill with all the words he wanted to say. Witty words. Words to make her laugh. Words to make her love him, as he loved her.

This could not go on; he just had to do something, but that something eluded him, and the cold winter slowly shivered on.

Christmas arrived, then Hogmanay, but still he could not speak to her and before he realised, the holidays were over, and he was back in school.

As the newness of the year wore away, and the snow slowly retreated up the glen, the days gradually grew longer. Then one unusually warm morning in February as he walked down the lane to catch the school bus, he noticed clumps of delicate white snowdrops growing on the roadside banks. He stopped and stared; then he saw a yellow crocus had emerged from the cold earth. It's almost Spring, he thought, and as he watched, a lone honeybee landed on the delicate petals. Busily it began to gather up the yellow pollen and pack it onto its back legs. Then the bleat of a newborn lamb drifted over the dry stone dyke from a nearby field and he realised the lambing had started. Spring, the time for love, he sadly thought. Then suddenly an idea entered his head. His face broke into a smile, but his thoughts were interrupted when his friend from the next croft appeared by his side. He turned a little embarrassed. The bees are out, he mumbled. A few minutes later, they were on the bus, but at last, he had the answer to his dilemma. Valentines Day was only a week away. He would send her a card, a card filled with his love for her. Over the next few days he spent hours composing a love letter, a letter that would put all that went before to shame. He filled the card he bought, with beautiful verse. When he read it through, he thought it was witty, funny and a little sentimental. He told her of his love and his longing, he compared her to the flowers, the sweet scented blossoms of spring. He wrote that she walked and talked like an angel. When he had at last finished, and the card could hold no more, he carefully sealed the envelope full of his unsigned verse, and lovingly sent it on its way. He just knew she would guess it came from him.

The Day arrived and as he wandered downstairs for breakfast, he wondered if she had opened her valentine's card. Perhaps she would call around this morning, and they could go for a walk. He glanced out of the stair window and saw the red post-van meandering along the lane. The conveyer of letters. The deliverer of his dreams. The guardian of his future happiness.

He entered the kitchen and saw his mother leaning against the sink with a broad smile on her face. She nodded towards the table, and when he glanced, he saw a letter propped up against the honey-jar. He picked it up, as if it were a delicate priceless object and read his name. He turned the envelope over and saw S.W.A.L.K. printed on the back. His heart began to pound and he felt his face flush. His smirking mother asked who his admirer was.

'Don't know,' he managed to reply with a red face and eternal hope in his heart. He carefully split open the missive, taking great care not to damage the priceless contents. He drew out a valentine card and saw that it was beautifully hand made. Made with care, made with love. The neat verse put his meagre efforts to shame. This author told of, HER undying love. And how she worshiped and loved him. One day, the author wrote, she would have the nerve to take him in her arms and never let him go. It was from her, he just knew it.

'Do you want anything from the shop?' he asked his mother trying to sound natural, but failing. He rushed his breakfast and quickly made his way down to the village. When he arrived, he slowly wandered passed her house, but she was not about, so he went to the shop and bought the tea bags for him mum and two macaroon bars, one bar for himself and one for his new love. Then as he slowly wandered towards her house again, he seemed to be having trouble with a shoelace. For every few steps, it seemed to need retying. Then suddenly he saw her. She came running out of her front door towards him. His heart raced as he awaited her soft embrace. Then she stopped and looked back towards her door.

A young man came out smiling, 'wait for me,' he shouted as he ran after her. The stranger took her hand and they skipped passed him as if he was not there. He stood with a smiled frozen on his lips and said 'hi.' But she drifted passed him in a dream, unaware of his presence. Her eyes were for the stranger. As he stood staring at her back, his heart broke into a thousand pieces. Suddenly tears welled in his eyes and ran down his cheeks. He wandered home confused and dejected. Why, did she send me the valentine card, he wondered. As he walked, he retreated behind his defences and swore that he would never again allow himself to love, never, never, never.

He spent a sad morning trying hard not to think about her, and gradually his heart hardened. Then his dog pushed open his bedroom door and sat at his feet looking up at him. She laid her head on his lap and whimpered. He stroked her head and she wagged her tail as she looked up into his eyes.

'Oh all right then, 'let's go for a walk.' His dog bounded from the room and he followed at a more sedate pace, more fitting to his new maturity. They went out the back door, across the field and into the birch wood. Soon he felt a little better as he and his only real friend climbed out onto the hill and looked down over the glen. They walked in a big circle and as they neared home, he saw her. His dog deserted him and ran to her. She dropped to her knees, ignoring the damp earth and played with and petted his dog. He slowly walked up to her and she stood to meet him. Her jeans were wet and muddy from the soft ground and her Barbour jacket hung loosely from her shoulders, revealing a warm home knitted jumper.

It was the girl from the next croft. The girl he had walked to school with since he was five years old. The girl he had always played with; but she was no longer a girl. For the first time he saw her, really saw her. Her looks put the policeman's daughter to shame. Her beauty was real, unabridged and complete. She leaned over and took his hand, and his walls of defences crumbled to dust.

Happy Valentines Day, she whispered.

End

Valentines Day.
Once called, Lupercian, and probably dates from an ancient pagan festival of love and fertility.

On February 15th, the Romans paid tribute to the she-wolf who suckled Romulus and Remus. They honoured their god of Nature and Fertility.

At the Lupercalia festivity's, young men ran around Rome, and lashed young women with strips of goatskin, in the belief that this could induce fertility.

In 496 AD, Pope Gelasius, outlawed this pagan festival and replace it with, 'to his way of thinking,' a more acceptable, Valentines day, using, Valentine was apparently martyred, for refusing to give up Christianity. He may have died on Feb 14th. 269 AD. He apparently left a note for the jailer's daughter, signed: - 'From your Valentine.'

S.W.A.L.K. (sealed with a loving kiss) Believed to date from World War Two.

Kissing under the mistletoe. That's part of a Druid fertility rite.

The Romans Visit to Scotland.
As I mentioned the Romans, here is a potted account of their brief visit to Scotland.

83 A.D. Agricola, the Roman governor of Britannia invaded southern Scotland.

According to Roman historian Tacitus, the Scots, known then as the Caledonians, 'turned to armed resistance on a large scale'. They employed guerrilla tactics. On one occasion, they attacked and almost wiped out the 9th legion. The Legion was apparently only saved when their cavalry rode to the rescue.

84 A.D. According to the Romans, the Caledonians met a well-equipped and organised Roman army at or near Inverurie. The Roman account states that during the battle they killed over 10,000 Caledonians. To my way of thinking that seems a far-fetched exaggeration. Shortly after the battle, the roman Emperor Domition, ordered Agricola back to Rome. The Romans retreated south. Again, this seems strange; if they had just won a major battle, why retreat. I suspect we will never learn the truth.

122 A.D. The Romans built Hadrian's Wall.

Later, Antoninus Pius, (86-161) invaded Scotland and built another wall in an attempt to push the frontier further north. This wall, between the rivers Forth and Clyde, did not last long. The Romans soon abandoned it and retreated south. Back behind Hadrian's Wall.

297 A.D. Scotland's Picts united; this was the birth of the Pictish nation.

306 AD. The Picts began to attack Hadrian's Wall and generally give the Roman forts on the border a hard time.

360 A.D. the Picts, together with the Gaels from Ireland, launched a successful and coordinated invasion across Hadrian's Wall into Roman occupied England. This marked the beginning of the end for the Romans in Briton.

On my way to School in Inverness, I sometimes stopped at a wee corner shop and bought my daily sweetie fix, usually a Macaroon bar. To make this delicious sweetie.

Macaroon bars.
14 g mashed tatties. Yes tatties.
450 g icing sugar
230g chocolate
115 g desiccated coconut.

As usual all amount are approximate. Mix the icing sugar into the leftover mashed tatties, a little at a time until the mixture becomes stiff and a wee bit sticky. Form into slightly flattened sausage shapes, place on icing sugar dusted greaseproof paper.

When they become stiff dip them in molten chocolate, and roll them in the coconut. that's it.

When a tried out this recipe I don't have the luxury of a big supermarket, so I rowed ashore and picket up a fallen coconut. I found an old bar of ordinary chocolate in the fridge; it was past its sell by date and a bit reluctant to melt, so I added a drip or two of water and a wee dram of rum, which worked well. The point I am trying to make is use recipes as a guide only, do your own thing.

Remember that in this recipe a little mashed tatties goes a long, long way. I think next time I'll add a few crushed nuts the mixture.

An Awful Man for the Drink.

When I first knew him, I thought he was an awful man for the drink. Not what you would call a drunkard, but he always seemed to have the smell of whisky about him. As to his age, well it was kinda hard to say, maybe seventy something; give or take ten years. I never knew him well at that time mind you, just enough to say hello and pass a few pleasantries with when we met by chance. He was always polite and never in a hurry. Even then, I was kinda drawn to this man, with his open friendly ways and his dry sense of humour.

I had taken early retirement and had bought a small cottage in the village. I went out of my way to try to sort-of, fit in. I must confess I was not having much luck with the rest of the locals. No one was rude to me mind, or said anything that I could take offence too, but most of them were, well a bit standoffish.

Then one summers day the man surprised me, the one I thought of as a drinking man that is. We met by chance; I was walking my dog on a quiet lane outside the village. He was sitting on a style, the one I had intended using; the one that led to the woods. He wore his usual long black coat and the green toorie with a red pompom that always sat on his head. The hat and coat were perhaps a bit out of place on that warm summer's day, as perhaps were his highly polished tackity boots.

His rugged features cracked into a smile when he saw me, and as I approached, he called out a greeting. We blathered about this and that for a while as he gently scratched my dog's ear. Suddenly he stopped talking, kinda leaned his head to one side and peered up at me.

'Do you like the whisky?' he asked in a very serious voice.

'Well now,' I said as I rubbed my chin, 'I always take a wee dram in the evening, just the one mind you, all things in moderation,' I added for his benefit. As a good Scotsman, I felt obliged to support the local industry. Besides 'the water of life,' to use the Gaelic translation, is a grand tonic, and always sets me up for a good nights sleep.

'Well in that case,' he said as he reached into the inner folds of his long black coat and withdrew a half bottle of the stuff. 'Take this home with you and have a wee dram on me.' He rose from his seat and pressed the bottle against my chest. 'Tell me next time we meet what you think o' it.'

With that, he cleared my and my impatient dog's path to the woods, and walked away along the fence line. I stood holding the accepted half bottle where he pressed it, and watched as he wandered up the gentle hill between the yellow whins out towards the torr. I don't know why, but I looked around and guiltily slipped the gift into my pocket. When I climbed the stile a moment later, I was deep in thought.

The pinewood was fairly big; it stretched for miles out over the hill and as far as the next glen. I walked along one of the many forest tracks. As my dog chased imaginary rabbits, I continued to ponder the source of the whisky. There were several distilleries in the area, and I wondered if it was sort of spare, and maybe given away free to the locals. Or, perhaps it was smuggled out of the distillery in the workers thermos flasks. 'Clearic,' I think they called it, and I had heard it was real firewater. Aye that's what it will be, I decided, 'Clearic.'

I walked on for a while wondering if I dare taste the stuff, when another thought suddenly struck me, so I stopped to consider this latest notion. Maybe he makes the stuff. Poteen, that's what it might be, Poteen, homemade whisky. I had heard stories about illicit stills hidden away in the hills but I had never seen one; or for that matter been offered a taste of the illicit product.

That evening in the security of my own home, I dubiously examined the bottle and its contents. As expected no label adorned the glass bottle and the substance within was almost clear. I unscrewed the stopper and poured a modest measure into my usual glass, held it up to the light and examined the liquid within. I could see now that it had a faint hint of gold, perhaps betraying its young age. I swirled it around and took a gentle sniff. This whisky smelt like no whisky I had ever known. I chanced a small taste, 'diluted for safety with a dash of water,' and to my great surprise, it was extremely good. Well more than that really, it was just like a very good mature malt, sort of sweet and smooth with a warm pleasant after glow. I even thought I could detect the faint hint of peat. Adding water certainly made it an exceptionally good dram, so I had a second wee glass just to be sure; it tasted even better.

I was eager to meet the supplier again, so next day I set out with an enthusiasm I had not felt in years. We met in the same place, and as before my dog ran up to him, no doubt expecting more ear scratching. I opened my mouth to greet him but before the words came out, he spoke.

'Well, did you like it?' he demanded as he stared at me.

I said that I did, but confessed that I added a little water as it was a little on the strong side for me.

'Just so,' he replied nodding, 'you'll not be used to the good stuff, being from the town. Now to business, how much would you be willing to give me for say a case, twelve bottles,' he added just to clarify the number. Well I certainly wanted his whisky, but I also wanted it for as little of my pension as possible.

'I like it just fine,' I started slowly. 'Has yon whisky got a name and where does it come from?' was my opening ploy to try for a good price.

'It's me that has it,' he replied, 'and you can call it what you like, so either make me an offer or no, it's up to yourself.' Realising that bartering was perhaps not his thing, I swallowed hard and offered him half what I was paying for my usual malt.

'Not at all,' he said standing up from his style seat, 'I'll no accept that.' Thinking that I had insulted him, I quickly added that that was only my first offer, 'sort of, to be discussed.'

'It's far too much,' he shocked me in reply. 'I'm no collecting tax for the government! I am just kinda sharing out a wee drop on the side to my friends. Now how about you giving me,' and he mentioned a sum that I felt embarrassed in accepting.

That was a good few years ago now, mind you, and I'm afraid to say that I became an awful man for the drink myself, and was soon one of his best customers. Then one day when I met him, I saw he was a bit on edge, a bit nervy maybe.

'Anything wrong?' I enquired.

'No, not really,' he replied looking at me in a funny sort of way, 'it's just that I might be wanting a wee bit of a hand with something, but I can't ask you because I don't really know you.'

'What do you mean,' I replied, feeling a bit insulted, 'we have known each other for over five years, ever since I came to the village.'

'Just so,' he said, nodding his head, 'I hardly ken you atall.' He then sat in silence for a while scratching at my dogs ear obviously deep in thought.

'I'll take a chance,' he suddenly said slapping his thigh and rising from the stile. 'Follow me.' He crossed the stile and headed away into the wood at a good pace, with me and my dog following behind. I was amazed at his fitness and had difficulty keeping up. Eventually he arrived at his destination and sat down on a log.

'Take the weight off your feet,' he suggested, so I sat down beside him.

'What is it you want a hand with?' I asked when I got my breath back.

'Patience man, all in good time,' then he went on to discuss old Malcolm. Malcolm had just passed away at the ripe age of ninety-three.

'We went to the school together,' he said, 'Malcolm was my pal, it's him that always gave me a hand when needed. Eventually he grew quiet and I realised he was greatly missing his departed friend.

We had walked right across the wood, and now sat at the edge looking out over the moor. It was a quiet day and I could hear the chatter of grouse. Far away, a pair of golden eagles soared and wheeled in the blue sky. A distillery, and a row of grey bonded warehouses lay on the glen floor, and I caught the faint whiff of fermenting barley on the rising air. Beyond the distillery, the river sparkled as it meandered its way towards the distant sea.

Half an hour passed by and the grouse grew quiet as the eagles neared our side of the glen. My companion suddenly rose and wandered back into the wood. He stopped for a moment and I could see he was listening carefully. A moment or too later he beckoned me to follow. I rose from my log and trailed after him as he followed a faint path. We entered a thicker part of the wood, and in the gloom, we wandered down a gentle slope and eventually came to a small clearing with an old stone dyke running through it.

A well cared for croft house, surrounded by old farm buildings, sat by a little used forest track. I saw a tidy vegetable garden. An old grey Ford tractor, apparently abandoned, sat surrounded by bits of ancient farm equipment. In one of the old buildings, I could see an old car.

I looked around at the pine trees, the distillery fence and the moor beyond the old croft as he unlocked the back door and stepped inside. Come away in, he said and sorry for the wee detour. It was to give me time to think. He wandered over to a half open door in the wall, when he opened it fully I saw an empty wood lined cupboard. He turned and looked at me as if suddenly undecided. I glanced towards the door as my dog came bounding in wagging his tail. When I turned back, the cupboard was gone. I looked into the darkness and saw stone steps that led down into the bowels of the earth. A match flared and I caught the whiff of whisky. A candle flickered into life and I saw my man beckoning me in. I stepped through, and carefully carried my dog down the steep stairway.

'Follow me,' he whispered after he had returned from closing and locking the doors. I did as instructed and followed as he led the way along a short passage. When we reached the room beyond, he lit a lamp; in its dim light, I saw his distillery for the first time. The room was about twenty feet by fifteen feet, with a smooth flagstone floor and roughly dressed sandstone walls. The vaulted, arched ceiling was black with soot with an opening that presumably led to the chimney of the old house. In the centre of the room, I saw the small onion shaped copper still. It sat over a pit hearth, the fire replaced by a gas burner. Oh, the still was a bonnie sight and the highly polished copper told of years of loving care.

A constant stream of clear water trickled from a pipe in the wall and splashed into an old stone sink. On a bench, made from flagstones, that occupied the remaining wall, sat lots of assorted bottles. On the other-side of the subterranean room lay six big wooden whisky barrels, next to them, a stack of staves and a few dozen steel hoops. A set of assorted cooper's tools hung on the wall.

He went over and carefully examined each barrel. When he was happy that all was well, I helped him roll them forward, he knocked out the stopper with a wooden mallet and water gushed out and ran away down a stone gully into a drain.

'Jimmy and I filled them a few weeks ago,' he informed me, 'just to be sure they were tight.' I saw a plastic hand pump with a red handle fixed to the wall, the type you sometimes see on boats to pump out their bilges.

'My father's old croft,' he said quietly,' this was his still, and he made the finest whisky in Scotland. I still have a few bottles left and I'll maybe let you have a wee taste one day. What time do you have now?' I told him it was just after midday.

'Good,' he said as he sat down and watched the water drain from the barrels. Ten minutes dragged passed to the sound of running water. When the barrels ran dry, I helped him roll them into the centre of the room where we chocked them, to hold them steady. He went over to the hand pump, fitted a clear plastic hose, then dropped the other end into a plastic bucket. He asked the time again and I glanced at my watch and told him. He nodded and sat down and scratched my dog's ear again. I could see he was deep in thought, and perhaps a little uncomfortable; I had no idea what he wanted a hand with, but was now quite content to wait and see. After ten minutes passed by, he turned and asked me to give the pump a few pulls. I tried working the handle but it was stiff and unyielding, I gave him a puzzled look.

'Give it a few more minutes then try again.' Six more times I tried the handle, then on the seventh try, it moved. He jumped up, and as I worked the pump, he held the hose over the bucket. Black liquid gushed and squirted from the pipe and I smelt whisky. When the liquid ran clear, he poked the hose into the first barrel. We took it in turns pumping the pump, and by the time the barrel was near full, the pump began to suck air. He moved the hose to the next barrel and twenty minutes later the process started all over again. It was a long hard day and by the time the last barrel was full, we were both exhausted.

'That's it,' he said as he laid a rag over the open hole and hammered in the last bung. 'We can have a dram then you can go home. I thank you for your help; I'll see you get a few extra bottles with your next order.'

Life in the village went on more or less as before. He had given me the use of a patch of ground for my own vegetable garden, so I had an excuse to drive my car into the wood. Now and then, he asked for a wee hand pumping whisky, carting bottled gas to the still house, or delivering cases of whisky to his various customers. The next few years crept past. I often wondered where the whisky came from, but never asked. Then one day when I just finished giving him another wee hand, and he had banged in the last stopper, we sat down for a rest.

'I have a wee surprise for you,' he said as he picked a bottle up from the shelf and placed two tumblers on the bench. 'This is the last bottle of my father's whisky and I would like to share it with you.'

We sat drinking the, 'amber nectar' and he talked about his past; he told me that he and his father once worked the land around here and rented the field the bonded warehouse now sits on.

'We had a wee digger business at one time; we mostly laid field drains and dug ditches. Well one day, the distillery manager offered us a fine wee contract. He wanted to put up new bonded warehouses and I suppose he felt a bit guilty taking back our rented field, so he asked us to dig the foundations and clear the ground. Well one day when we were digging, we came across the copper water pipe that once fed a field trough. Now we knew the water in that pipe came from the well above the still house, and we realized the importance of this discovery. So during the construction we fed a water pipe up inside all the buildings and fitted them with a tap. When the warehouses were complete, we turned off the water at the well, cut the pipe and fitted a pump. The taps were, as expected just forgotten about. Now, once in a while, the distillery handyman, who just happens to be a cousin of mine, goes around the bonded warehouses and carries out any maintenance that may be required. An excise man always accompanies him, just to make sure he doesn't accidentally carry away the odd barrel of whisky. Now the excise men are lazy, and they just sit on a chair at the door while the work is going on. Once out of sight of the guard, my cousin fits a hose to the tap and prizes the bung out of a barrel and we pump out the whisky. Then over the next few months, I run the stuff through my still a few times to polish it.

'But won't they miss it?' I asked.

'Not at all, barrels are always leaking and the stuff has no value anyway, it's the government tax that makes whisky expensive. Anyway my cousin is retiring next week, and I'm getting on a bit, so this may well be the last drop we get.'

Then just a week later, the old man was playing in a bonspiel (curling match) on the deer pond, when he slipped on the ice, landed on a curling stone and broke his hip.

That's it I thought no more of his whisky for me. I counted out my supplies and decided that with care I could last out for about a year. Then it would be back to the off-license. He got over his fall

all right, but felt he was no longer fit to curl or work at his we sideline. Then about six months later, when I was visiting him, his cousin arrived, you know, the one who used to be the distillery handy man. Well a bottle of the good stuff appeared and the cork thrown into the fire and we got to talking about whisky.

'Now then,' the cousin said,' I have a wee bit of news for you both, my boy Ted has got himself a new job, my old job.' This news went down well and over the next few months, I received an intensive course in the manufacture of whisky and quickly learnt that it was an art rather than an exact science. I learnt the importance of temperature and new words like foreshots and aftershots.

The events mentioned above took place twenty years ago, and the whisky I now refine is I think, just about as good as my teachers. Sadly, my teacher died last month, it was on his one hundred and tenth birthday. In his will, he left me the title to his croft and his old black coat and toorie. (Beret type hat with a pom-pom)

'Now how much will you give me for say, a dozen bottles?'

End

Whisky, from the Gaelic,' Uisce Beatha' has played an important part in the life of Scotland's people for well over a thousand years..

As to its origins, they are lost in time, but several theories abound. One suggests that the art of 'distilling,' originated in the Middle East around eight-century AD. Perhaps, Christian monks brought it to Scotland and Ireland. Who knows? Even the name of St Patrick, has been evoked, as a peddler of the new technology. But wherever the original knowledge came from, the Scots have evolved and adapted the technology into an art form, and now, as we all know, we produce the finest whisky in the world. What is known for certain is that the Ancient Celts made the stuff.

Curling.
The old man was a curling fan, yes and like golf, it's an old Scottish game. Now curling is a very competitive sport. Especially when played indoors, but it can, and is played, by people of all ages.

The out door game is usually more relaxing with the odd dram, and sometimes someone will arrive with a big pan of stovies.

Nairn Curlers club on Ardclach pond.

Atholl Brose. is an ancient Highland drink, first recorded in 1475. There are several legends associated with it. One states that the Earl of Atholl, who was attempting to capture, Iain MacDonald known as The Lord of the Isles, put the mixture into a well. The lord, who was fond of the tipple, stayed at the well too long and was captured.

This is the traditional recipe made public by the Duke of Atholl some years ago.

Atholl Brose.
3 rounded tablespoons of medium oatmeal.
2 tablespoons heather honey.
Whisky.

The oatmeal is prepared by putting it into a basin and mixing it with cold water until the consistency is that of a thick paste. Leave for half an hour, or longer, then put through a fine strainer, pressing with a wooden spoon to extract as much liquid as possible. Throw the used meal to the hens and use the creamy liquor from the oatmeal for the brose. Mix four dessert-spoonfuls of pure honey and four sherry glassfuls of the prepared oatmeal and stir well. Put into a quart bottle and fill with malt whisky;

shake before serving. An alternative is to add cream, this will give you a drink similar the 'Baileys Irish Cream.'

Hot Toddy.

It's a medicine, and is almost worth getting the flu so you can sample it. But the Scots are a canny lot and normally take it as a preventive, or just in case.

Put one large dram into a glass or coffee mug, add a teaspoon of heather honey, top up with hot water, then go to bed and sip your ailment away. If you have no heather honey, you can use any honey, or just add sugar.

Whisky Butter.
100g butter.
100g soft brown sugar.
1 dram of whisky.

Cream the butter, add sugar and then add the whisky a little at a time. Nice on Christmas pudding.

Whisky coffee or tea.
Out on the hill red deer stalking, or sitting by a salmon river, add a wee dram from your hip flask to your coffee or tea; it sort of spices it up a bit.

Oats and Oatmeal.
Was and perhaps still is an important staple food in the Highlands. Many people still have porridge for breakfast, I do. It is also one of the very few foods that a seasick sailor can 'keep down' during a storm at sea.

Old insult, perhaps.
A light-hearted insult to a Highlander is to suggest that he was (a bowl of meal), - Fraser, Grant or whatever his clan name. In days long gone, if a Chief wanted to increase his clan numbers, he would offer a bowl of meal, to all who joined his Clan and took up the clan name. Not to be taken literally, but I suspect meal played an important part of the inducement.

This next recipe is from the island of Barra. They used to make Oatmeal Bannocks. from the first grain of the year, to be eaten on St Michael's day. (29th of September.)

Oatmeal Bannocks
100g self-raising flour.
1 tsp' bicarbonate of soda.
1 tsp' cream of tarter.
100g pinhead oatmeal.
1-tablespoon honey or syrup.
About (3 fl oz) milk.
2 eggs.
A pinch of salt.

Sift the dry ingredients into a bowl; mix in oatmeal, then the eggs and milk to form a thick cream. Cook on a hot greased girdle, or heavy frying pan. They should brown and rise. When cooked, put onto a wire rack to cool. Wrap in a tea towel and keep warm until ready to serve. They are a bit like pancakes, but with more body and taste, not sweet and a good recipe. I enjoy them with honey or golden syrup.

The Exile.

When William Fraser left the Highlands of Scotland, he did so under a wee bit of a cloud.

As a young man, he was perhaps a bit on the wild side and several times in his youth, he had stood before the local magistrate and promised to mend his ways. He had never been content to work on the family farm; he wanted more out of life. He was in fact a bit of a disappointment to his father who always hoped that his only son would take over the tenancy of the farm; but this was not to be.

Then one day the army-recruiting sergeant visited the village, and William, strongly advised by the magistrate, accepted the offered shilling. He marched away full of confidence, his head filled with the sergeant's stories of excitement and adventure. But when he reached the barracks and his basic training begun, he discovered that the army was not quite, what he expected. He soon realised that joining up may have been a mistake. Give him his due though, he did stick with it for a while, that is until a weedy, toffee nosed idiot of an officer took charge.

Ordering a big powerfully built youth like William to gather up and paint wee stones white, and then place them all around said officer's quarters was bad enough. But, when the frustrated officer's wife asked him to help her shift a wee bit of furniture; that just turned out to be in her bedroom and then when the lady suggested that they might just test the bedsprings. Well what was he to do, she was an officers wife, and therefore had to be obeyed. She was certainly bonny enough with a fine slim figure and William was far from slow on the uptake. Anyway I'm quite sure you wish to be spared the details so I'll just say that over the next few months, the lady seemed to have an awful lot of furniture that needed shifting.

Well all good things must end, and this little diversion was no exception. It happened suddenly and without warning. It literally caught William with his trousers down. William and the lady were, shall we say 'shifting furniture again,' and making a fair noise about it, with lots of …, sorry I forgot I was going to spare you the details. Anyway, the officer had come home for something or other, and hearing the commotion in the bedroom, went to investigate. To say he got a bit of a surprise is perhaps an understatement. It must be noted however, that this officer was well known, for his total lack of understanding towards the common soldier. Being caught in bed with the officer's wife was hardly a hanging offence. William however was not too hot on military law, and decided it might be better if he just left. So he gathered up his kit and climbed over the fence. He was standing by the roadside gathering his thoughts when a wagon appeared so he hailed the driver and cadged a lift into town. He sold his kit to a man in the pub. When I say his kit, the army did give it to him, but he dumped his uniform, as no one wanted to buy it. I suppose that's quite understandable when they give new ones away free.

He then headed for home, back to his fathers farm above Loch Ness. His father was glad to see him, well that was until he mentioned that he had left the army unofficial sort of, and they might just come looking for him. He borrowed fifty pounds from his father and headed south.

Some time ago, he had read that Australia was full of gold, and all a man had to do was pick it up, so he decided he would go to Australia. He headed for Glasgow and looked for a ship going south. He had lots of money, courtesy of his father and could well afford to buy a ticket, but money was money, so why spend it, if there was no need. He asked around in the dock and harbour pubs, and soon discovered that by signing on as crew, he could get to Australia for free. And get fed and paid into the bargain. So he did the sensible thing and signed on.

The ship was not due to leave port for several days, so William had time to explore his new home and learn the ropes. One day he saw a squad of soldiers marching along the harbour side and assumed they were looking for him, so he hid below decks. He was glad he had taken the precaution of slightly changing his name; he was no longer William, now he was Bill.

When the ship finally left port, Bill sighed a sigh of relief. He had escaped and become a sailor. He worked hard at his new profession, winding big windlasses or hauling on one of the jungle of ropes. Bill was a strong man and became very popular with the crew who liked to have him on their watch, less work for them of course. Bill could pull two men's weight with ease and three if he got angry. During the long voyage, he became friends with another man, a fellow sailor who was also on his way to the gold fields, so they teamed up. The other man said he knew all about gold prospecting and Bill had almost fifty pounds.

They sailed south around the Cape of Africa, weathered the usual storms, and eventually landed in the small settlement of Fremantle on the west coast of Australia. From the deck of the ship, they saw a land that was flat and dry, with the smoke of bush fires climbing high into the air obscuring the horizon. Bill and his friend were amazed, and a little shocked by the number of people already in this supposedly empty land. They deserted the ship and walked towards the town. On the way, they passed camps filled with thousands of men, from dozens of different countries. Soldiers on horseback rode past, leaving clouds of dry dust in their wake. In the town, they were accosted by men, who tried to sell them gold mines, or rich claims. They passed by brothels and drinking houses, with their customers spilling out onto the street. They found a store selling mining supplies and joined the queue of eager immigrants hoping to strike it rich. When they reached the counter, they bought a miners kit, consisting of two gold pans, one pick and two shovels a hand compass and map. They also bought broad brimmed hats more suited to the hot climate, and some basic camping and cooking equipment. Horses were scarce and those available were extremely expensive and well beyond their means. They bought an old rifle and ammunition from a drunk in the street. Then the next day they headed inland. Bill and his partner made frequent diversions into the bush on either side of the trail to hunt and do a bit of prospecting. But the gold eluded them so they marched on. Several weeks later, as they neared Boulder, they heard the noise of the buzzing town drifting out over the bush. They had arrived at a wild boomtown. They rested up for a few days to gather information. Most of the people doing the talking didn't really know, and those who did know, remained quiet. But they got the general picture. Water was, as they already knew, the biggest problem. The bars kept whisky on the counter but the water stayed safely hidden. Not only was there no water to pan the soil for gold, but no, or very little water to drink. It was available of course but like all scarce commodities, it was expensive.

They marched out into the bush for three or four days, then they would look for gold. Some times, they were lucky and found a small nugget just lying on the surface. Other times they went for days and found nothing. They shot kangaroos for meat and quickly discovered that the tail was the only bit worth eating.

Several months passed by, then one day when they were ten days out of Boulder, they found the dried up body of a man, with his mining kit spread around him. Among his stuff, they found a bag of gold nuggets that weighed about twenty pounds. They decided that the rich man probably died of thirst. They buried the man and hid his gold, then set out to backtrack his footsteps. Two days later, they lost the trail, their water was low and they wisely decided it would be foolish to continue. Bill shot a kangaroo, and in the cool of the evening as they sat around the fire cooking, they decided that in the morning they would head back to Boulder, sell the gold and return better prepared.

Bill suddenly glanced up and saw a tall Aboriginal man with grey curly hair; he carried a long white spear that must have been ten feet long.

The black man was very thin. Bill could see the outlines of his bones. The lone bushman stood very still on one leg, and appeared to be looking at the kangaroo tail that hung from a tree

Bill beckoned the native Australian to join them. The man smiled a flash of white teeth, and then slowly came forward, with his suspicious eyes dancing this way and that as he approached.

When he reached the campfire, he sat on his haunches and spoke to them. Bill spoke back, neither understanding the other. Bill offered the stranger some meat. The man smiled broadly, stood up and shouted towards the bush. Suddenly the trees and scrub came alive with moving figures.

'We are going to need another roo'.' Bill said as twelve hungry people gathered around their fire and quickly demolished the kangaroo tail. They asked in sign where the rest of the beast was, so Bill wandered away into the bush followed by half the new company. He found the remains and they dragged it back to the fire. Over the next hour, Bill watched the tribe consume the rest of the kangaroo and discovered that the native Australians preferred their meat almost raw. As the night wore on one of the old men signed that he wanted a drink of water. This was difficult, but Bill being a Highlander could not refuse. He carefully poured out half a mug of the precious water and handed it to the old man who smiled, washed out his mouth and then spat the water out onto the dry earth. The look on Bills face made the new company burst out laughing and Bill felt that he had been set up for this foolish pointless joke.

In the morning when he awoke, the company was gone, and later as the men sat around the fire eating their breakfast they discussed their visitors.

'Well,' the partner said, 'what do you make of that pointless joke with the water.' Bill remained quiet for a few moments deep in thought, and when he spoke his voice-betrayed excitement and his eyes sparkled.

'That was no joke they played on us last night,' he said.

'They know the value of water and they would never waste it.'

'Unless,' the partner said realising the significance of the statement,' there is water right here.' It took them all morning to find the water; it bubbled out of the red earth, travelled a few yards and then disappeared back into the ground. This is our new camp

Bill said smiling. Over the next month, they scoured the surrounding land until at last they found the gold field. The gold lay on the surface, they picked it up, filled their hands and pockets. A week later, Bill, left the partner on guard and set out on the long walk back to Boulder to file their claim.

Bill and his partner soon became among the richest men in Australia, and when they had gathered up all the surface gold, they sank a mine, and followed the yellow seam down into the bowels of the earth. Within a year, they employed two hundred miners, and the surrounding lands swarmed with prospectors and new mines.

Bill built a big house in the new settlement that had sprung up and sat back to enjoy his wealth, but with little to do now he began to think of his real home in Scotland. He thought about the hills and the glens, and he thought about the silver salmon and the red deer. He longed for the taste of haggis and neeps, and the skirl of the pipes. He desperately wanted to go home, but he knew he could not. For a time he tried to find comfort in drink but found only more sadness. Then one day he sent his father most of his remaining fortune, loaded up a wagon and went prospecting in the outback. But luck eluded him and he found no more gold. He wandered for a year and a half and when he eventually turned back towards his house, it was with a feeling of loneliness and the burden of failure. What is the point of wealth he thought if he could not spend his days in the country he loved. As he reached home, a deep depression settled over him. He opened his door and unbuckled his holster, the weight of the revolver-hung heavy in his hand.

'William is that you,' a familiar voice rang out. He lifted his head and saw his father before him. His fathers embrace and the news of his pardon brought tears of joy to his eyes.

End

This story is fiction, but it could well be true, for thousands of men from the Scottish hills and glens went out to Australia in search of yellow gold, and a few even struck it rich. My own ancestor, 'William Fraser,' 'late of Balnain' left his father's farm in the early

1800s, and wrote letters home about his adventures. He described the sailing ship that took him south and the hard back breaking work of sinking shafts, and diverting rivers. He wrote about the cost of food and the price of a new pair of boots.

When I visited the workings, I was truly amazed by the amount of work the miners carried out. Burns and streams were dug down to bedrock, some to a depth of over 30 metres, '500 plus metres wide at the top,' and they followed them for miles. Shafts in excess of 30 metres deep were common, and I saw square miles of flat bush that the miners had dug down to the bedrock and trenched.

The lucky ones, like Bill and the partner found nuggets lying on the ground. I don't know if my William struck it rich or not, but I hope he did.

When I crossed the Australian Nullarbor desert, in search of my lost cousins, three thing's stand out in my mind. First the billions of flies, second the Corbies, (birds of the crow family,) and thirdly the bush fires that always seemed to be on the horizon.

The thought of the flies still makes me shudder, for they literally got everywhere, and 'going' behind a tree, or putting up out tent became a near nightmare experience.

As for the Australian black and white corbies, 'whose lilting song I just love,' it was so hot and dry that they stood around under trees looking dejected; their beaks agape and almost unable to draw breath.

The speed the fires travel at is truly scary, we had one close encounter; one was more than enough.

Recipe. Just for fun.
Now have you ever tried 'Kangaroo meat,' there are numerous recipes for this healthy low fat and versatile meat. Go on have a taste, 'it's now available is some supermarkets,' you will be surprised how tender and succulent it is.

A few cooking rules for Roo meat. First, unlike venison, you do not need to hang it, and you can eat it fresh. Due to its dryness, it's always best to brown the meat, (all sides) in a sizzling hot pan.

Never add salt, and do not over cook, in fact rare to medium, rare is always best.

Roasts; wrap in foil and cook in a preheated oven (220°c) for about 10 min per pound.

Tail steaks; pan fry in a little oil, about two minutes per side. Or try the burgers; honest, they really are very good.

Right, back to Scotland, so how about, Haggis. It's OK, I know fine there is no way in this world you would ever dream of making the haggis like my ancestors did. I don't blame you atall, (at all) but there are a few very good alternative recipes. But first:-

Burns Night Party, 25th, January.
In December, or a month before your Burns Night Party, begin to learn the words.

To A Haggis.

By Robby Burns.

Fair fa' your honest, sonsie face,
Great chieftain o' the puddin'-race!
Aboon them a' ye tak your place,
Painch, tripe, or thairm:
Weel are ye wordy of a grace,
As lang's my arm

The groaning trencher there ye fill,
Your hurdies like a distant hill,
Your pin wad help to mend a mill
In time o need,
While thro your pores the dews distil
Like amber bead.

His knife see rustic Labour dight,
An cut you up wi ready slight,
Trenching your gushing entrails bright,
Like onie ditch;
And then, O what a glorious sight,
Warm-reekin, rich!

Then, horn for horn, they stretch an strive:
Deil tak the hindmost, on they drive,
Till a' their weel-swall'd kytes belyve
Are bent like drums;
The auld Guidman, maist like to rive,
'Bethankit' hums.

Is there that owre his French ragout,
Or olio that wad staw a sow,
Or fricassee wad mak her spew
Wi perfect sconner,
Looks down wi sneering, scornfu view
On sic a dinner?

Poor devil! See him owre his trash,
As feckless as a wither'd rash,
His spindle shank a guid whip-lash,
His nieve a nit:
Thro bloody flood or field to dash,
O' how unfit!

But mark the Rustic, haggis-fed,
The trembling earth resounds his tread,
Clap in his walie nieve a blade,
He'll make it whissle;
An legs an arms, an heads will sned,
Like taps o thrissle.

Ye Pow'rs, wha mak mankind your care,
And dish them out their bill o fare,
Auld Scotland wants nae skinking ware
That jaups in luggies:
But, if ye wish her gratefu prayer,
Gie her a Haggis!

Aye, you're right, it is a bit difficult to pronounce the words, but learning a verse or two is well worth the effort.

The day before your Burns Night Party. Make your

Recipe Haggis, old and new

Up to 200g of liver, I use about 100g
225g minced beef. If your butcher sells heart, use about.
100g and bit less mince.
2 onions.
180g oatmeal. (Porridge oats pulsed in the blender)
A little Olive oil.
OK you want to know how much, about a tablespoon full.
1-teaspoon salt.
1-teaspoon pepper.
More if you and your dinner guests like the stuff.
A little stock or the water you used to boil the liver.

OK, that's the list of ingredients, and it will give you, a good tasty and passable Haggis.

Boil the liver for about five minutes, take it out, and leave to cool. Place oatmeal in 'dry' fry pan, and roast until it turns light brown. Now comes the hard bit, peel and chop the onions, chop the liver up as finely as you can. Now using your hands mix the whole lot together. Add a little of the stock until the mixture is just moist, don't make too wet. Put it in a clean cheesecloth bag, or your old pillowcase, place in a basin, cover with greaseproof paper or an upside down plate, then steam for two hours in your pressure cooker. When haggis cool sew it up.

Then one hour before the supper, return it to the steamer.

Ply your guests with good whisky, then slip out and put on a CD with bagpipe music, and carry in your steaming haggis. Recite 'To a haggis' and plunge a knife into the 'chief of the pudding race,' serve with tatties and neeps.

If you can't find neeps, butternut squash works OK but add a little pepper to it. Good luck and have fun.

Traditional Haggis.

Since you were wondering, here's an old recipe for Haggis. You will need the bits and pieces from inside a sheep.

Ancient Haggis recipe
The stomach bag.
It's liver.
The lites. (Lungs)
The heart.
1 cup oatmeal.
2 onions.
200g suet.
Salt / pepper.

Clean the stomach bag then leave soak it overnight in salted water. In the morning, turn it inside out. Put the heart, lights, and liver into a pan. Bring to the boil. Simmer for 1.5 hours. Toast oatmeal.

When cool chop up the heart, lights and liver. Mix the lot together with the suet, add salt and pepper. Keep moist using the water you used for the boiling. Fill the stomach bag just over half-full with the mixture then sew it up; you don't want it to escape.

Put it in a large pan of water. When the haggis begins to swell, prick with a skewer to prevent it bursting. Boil for 3 hours.

Sounds gross, but it tastes wonderful. Remember that in days gone by, our ancestors could not afford to waste food, so nothing was wasted.

Wild rabbit is delicious, can be cooked as if it was chicken, you won't be able to tell the difference.

Rabbit Hotpot
1 chopped up rabbit
3 large onions
A few carrots or any other vegetable you have handy.
Add your favourite spices or herbs.

Flour the meat in seasoned flour, place onions and meat in layers in the dish, onion on the bottom and top If you have bacon handy, place some under the top layer of onions. No water required Bake for about 2 hours.

The Bell.

Somewhere in the distance, a bell ring.

The white haired old woman lay in her hospital bed with a far away look on her wrinkled weather beaten face. Her dim watery eyes seemed to be staring at some far away unseen object. Meanwhile her thin bony hands were moving back and fore, performing a task known only to some remote part of her aged confused mind.

The sight tore the heart out of her loved one who stood by helplessly looking on. It was Lindsey who stood by her grandmother's bed; her hands tightly gripping the bunch of fresh flowers she had lovingly picked from her garden. But now the stems bent and snapped under her tight grip as a tear slowly trickled down her cheek

Her grandmother had always been a tower of strength, quick witted and extremely fit for her eighty-eight years. Now suddenly it had all changed. Her grandmother appeared to have shrunk. She lay as an empty chrysalis case after the butterfly had flown. The nurse came over and gently removed the flowers from Lindsey's crushing grip.

'She doesn't know me,' Lindsey whispered as another tear squeezed out from her moist eyes and followed its companion to the floor.

The old woman was unaware of her granddaughter's sadness; in fact, she was far too young to have a Granddaughter. She was young care free and happy. She was a child again; except she did not know it was again.

She was playing in the field behind her parent's cottage. In her mind, she saw the blue Scottish mountains and the dark sparkling waters of Loch Ness. She sat among soft meadow grass wearing a cotton floral frock that surrounded her like a doily. She dropped her eyes from the majestic vista and concentrated on the daisy chain she was making. Her tongue appeared between her lips as she carefully pushed her thumbnail through the stem to make an eye. Then with great concentration on her face, threaded the next daisy onto the chain then lifted her old arms high and gently placed the garland on her little sister's head. It slipped over her siblings face and hung down over one eye. The old lady made a dry cackling noise, and the young girl laughed, giggled and waved her arms in the air; what fun. Then she smiled happily, as she reached out and rearranged the daisy chain.

'Look Frances' the nurse said as she placed the vase of flowers on the bedside cabinet, aren't they lovely, your granddaughter brought them. Do you want to sit up?

Frances looked up and saw her father walking towards them leading his huge grey Clydesdale horse. Do you want to ride home on Dobbin, he asked. Frances nodded her head enthusiastically and held out her arms towards her father, then felt his strong arms gently lifting her.

Then for a fleeting moment a strange image came into her head, two people were by her side, one dressed in white, but the image dissolved as her father placed her on the broad back of the huge horse. She quickly gripped the coarse hair that run down the powerful neck and entwined her fingers in the soft bed blanket. This was a real treat and she sat proudly with her little sister between her arms. The back of the horse was so wide and so high; it was like riding on a magic carpet. She sat propped up with her pillows turning her head this way and that, hoping that some of the other kids in the village could see her. From her elevated vantage point, she could see all the way down Loch Ness. Billowing smoke puffed out of the steamer's funnel and drifted up into the clear blue sky. She let go one hand and waved a weak arm at the Gondolier, and far - far away she heard the hoot of its steam whistle.

Again, she felt the strange tugging and thought she was falling, but she gripped tightly with both hands. Then she was home and her mother was scolding her father for letting them ride home on

Dobbin. Her father was laughing and smiling as he lifted her and her sister from the horses back. She ran towards the back door scattering the hens as she dashed into the kitchen. Suddenly the smell of fresh bread filled her senses, she stopped, sniffed deeply, savouring the wonderful aroma. Then a moment later, as the nurse led her distraught granddaughter away, the old lady began to snore quietly.

Close by, as a bell rang, a wave of confusion washed over Frances. She lifted her head, and with a concerned look on her face saw Mr Campbell, the schoolmaster standing by the school door. He was wearing his shiny baggy black suit, and she laughed when she saw his baldhead reflecting the sunlight.

He was standing in front of the door with a gleaming brass-bell in his hand, and as she tilted her head to one side, she heard the schoolchildren chant a rhyme. Her silent lips played homage to the song that she had sung over eighty years ago.

Our wee School
(Anon)

Our wee school is the best wee school,
It's made o' bricks and plaster.
The only thing that's wrong we it,
Is the baldly headed master.
He goes to the pub on Saturday,
He goes to church on Sunday.
To pray to God to gi'e him strength,
To murder the weans—on--- Mon----d---a----y-----

The singing voices drifted away into the distance, only the sound of the bell remained in her head. Her lips stopped moving and she lay back against her pillows. Slowly the ringing faded into the distance and it grew still and quiet.

'Frances are you awake' the nurse asked. 'Frances, are you all right?'

Suddenly she sat up straight, she was in the big schoolroom, her black slate with a wooden surround lay on her desk and up on the black board the copper plate writing was waiting for her to copy.

Mary, her best friend sat by her side at the double desk, she was whispering something. The old lady leaned closer to hear the words but Mary had finished talking. She heard a sudden loud noise and looked up. One of the big boys, Jimmy Rose was standing in front of the class with his hand held out and Mr Campbell the master stood towering over him. Hold your hand steady boy, she heard him shout. She felt afraid as she lifted her wrinkled fist to her mouth and bit on her knuckles. She heard the swish of the black leather belt then shuddered when she heard the smack as it made contact with Jimmy's hand. She screwed her eyes up tight but again and again, the noise came, each time louder and louder, and as her fear grew, her body began to shake and tremble.

'This should settle her,' the doctor said as he withdrew a needle from the thin drip tube that led to her thin and wrinkled arm. The nurse nodded and gently tucked in the blanket, then stood by for a while as her patient slipped away into a restful sleep.

The bell was ringing again, it was playtime, and she was skipping with the other girls and chanting a rhyme. A puzzled look crossed her face as she struggled to hear the words. She got annoyed and for a moment, her face contorted into a scowl. Then the words broke through the haze and her face relaxed and she smiled as the words filled her head.

> *One potato, two potato, three potato, four.*
> *Five potato, six potato, seven potato, more---*
> (Anon)

The skipping rope ceased to slap on the paved playground, and the skipping boots fell silent. The chanting of her playmates continued for a few moments then slowly faded away. A strange voice was intruding, penetrating, it was very insistent.

'What's that?' Frances said, trying to answer the annoying voice in her ear.

'It's me, Granny,' the voice said. Her gaze shifted on to the person by her bed and she tried to focus, then slowly a face appeared before her.

'Oh, it's you Heather, I was just dozing, what is it.? Oh! Have you been crying?'

'It's Lindsey, Granny; I just popped in to see if you wanted anything.'

'I'm fine,' Lindsey's grandmother said in a confused voice as she studied her visitor, 'just fine.' She looked around, 'am I in the home now?'

'No Granny, not yet,' Lindsey replied taking her grandmothers hand.

The bell sounded to mark the end of visiting time and Lindsey gently kissed her sleeping Grandmother and tiptoed away.

Frances heard the bell sound, and sensed she was alone. She felt the gentle kiss, and a thrill passed through her young body as she pulled back, don't Jimmy, father might see, she said the words but did not mean them. Then Jimmy was standing before her in his smart new soldier's uniform.

She heard a bell ring and saw the white ambulance with a big 'Red-Cross' painted on its side. Suddenly a loud explosion ripped through the air and it went dark. An air raid siren howled a belated warning. As she stood listening, the big ark spotlight suddenly burst into life emitting a loud buzz. She saw the rubble on the street, heard the exploding bombs, and saw bright flashes that filled the sky. The sound of Ack-ack fire rattled all around her and tracer bullets streaked out across the night sky. She smelt the toxic smoke, and much worse that filled the air. Mary was standing by her side and they were steering the big spotlight trying desperately to find the enemy aircraft that flew overhead.

Suddenly the smoke grew thicker. She could hardly catch her breath. The ground under her feet began to shake and she felt afraid. The pounding of the bombs grew louder and filled her head. A wave of pain washed over and through her. Then she felt the prickle of the stiff uniform on her arm and something pressing onto her face. As she struggled, the sound of exploding bombs faded away. As the pain left her, she became drowsy. A moment later, she was fast asleep.

As the doctor dropped the used syringe into the dish, he looked at the nurse and shook his head. Then as he walked away, the nurse adjusted Frances's oxygen mask. She held the old lady's hand and watched as she relaxed. When her breathing steadied, she picked up the kidney dish and walked out to call Lindsey on the telephone.

The bell sounded again and a voice was calling her but she could hardly hear it above the sound of the bombs, she turned her head towards it and the voice became louder.

'Granny, can you hear me?'

The old lady opened her eyes. 'Of course I can hear you, who is it?'

'It's me Granny.' Lindsey smiled as she recognized the spark of life returning to the grandmother she dearly loved.

'How are you today?'

'I'm fine Lindsey; just fine, did you see Mary she was just here?'

'No, sorry Granny she must have left.' Lindsey replied as she looked around wondering who Mary was.

'Not to worry,' her grandmother said glancing towards the door.

'The nurse told me you are doing fine.' Lindsey said with a smile.

'Is that so,' the patient replied, 'and how would she know, I never see her.' Over the next few days, Lindsey's grandmother lost touch with her past as the hospital staff nursed, bathed, and fed her. She told Lindsey she was able to get up and walk around slowly with one of these things old people use, trimmers or Zimmer's or some such nonsense. The weeks slowly passed by and Lindsey thought that her grandmother was well on the road to recovery.

Then one day Frances heard the bell again; people were dancing in the street and the church bells were ringing. Everyone was happy, oh so happy! The war was at last over; she was standing by the dock watching a big ship as it came over the horizon. People were cheering and waving and then as the big ship drew closer she saw soldiers lining the deck and she jumped up and down in excitement. The ship blew her whistle as she called out his name.

Suddenly Jimmy was standing in front of her holding her shoulders and looking into her eyes.

'Jimmy, oh Jimmy, is it really you, I have been getting a bit mixed up recently.'

'It's me darling, I have come home to you,' she felt a tear trickle down her cheek as she tightly held on to her Jimmy, she was so happy.

'Oh Jimmy, I have missed you so much.' She muttered the words over and over again, as the nurse patiently held her hand.

The church bell was ringing; she was walking down the aisle with her father. A moment later, she stood before the altar with Jimmy. She wore her long white wedding dress. Her sister was there, so was Mary, and all her friends from the village. She held tightly on to Jimmy as he slipped their wedding ring onto her finger.

The bell rang and she felt the pain of childbirth followed by a wonderful feeling of happiness as she looked down at the tiny child in her arms, and she heard a clear voice say, I baptize thee. The bell rang again and a voice was calling out to her.

'The letter came,' it said, she looked up and saw Jimmy running in calling her name in excitement.

'Frances, Frances, look what I have.' He was holding up a letter and waving it in the air.' We have been accepted,' he said,' we can go.' They stood by the railing of the big ship that was taking them to their new life far far away, and as she looked out over a wide and empty sea, she felt faint. The reassuring presence of Jimmy's arm around her made her smile and she saw him standing by her side holding Heather in his arms. Then as the vast new land came into sight, a bell rang. Suddenly she was standing at a stone sink with the pump handle in her hand, and she screamed out in pain.

Then Jimmy was holding her in his arms and she relaxed. He was telling her he loved her, and that they have a son. Then she felt him kiss her cheek, but the kiss turned wet as she saw the tangle of twisted steel that was once her son's car. Her tears filled her eyes and she quietly sobbed. A bell was ringing and the tears became a flood and ran down her face, as she stood by her son's grave.

Heather and Jimmy were by her side, Heather had her arm around her shoulder, and she heard her say something but failed to catch the words.

The bells were ringing again and the church was full of happy chatter, the organ was playing here comes the Bride, then the sound changed to the sounds of children chanting.

Here comes the bride,
Sixty inches wide.
See how she wiggles her big backside-
(Anon)

She got very agitated, and waved her arms and shouted out to chase the children away, then the image slowly cleared and once again, she was in the church. She looked around and saw Heather dressed in white. Jimmy was walking proudly by her side.

The bell rang, and Heather and her husband were standing on the doorstep of her and Jimmy's house with a child in her arms. Hello Granny she said, this is your granddaughter Lindsey.

She was now walking through a wood with Heather and her husband. Lindsey was running ahead with her pigtails flapping, and Sally the spaniel danced at her feet. She looked around for Jimmy but he was not there, and she felt grief as a sudden pain invaded her chest. A distant bell sounded and she thought she could hear the sounds of running feet. Then she saw Jimmy standing by her side.

'Come my love,' he whispered, 'it's time.' She gladly reached out towards her husband, and as their hands met, her pain faded. She rose from her bed as a lifetime of wear and tear drifted away like mist on the breeze. She stepped into her husband's arms, and together they wandered across the field towards the loch.

The bell rang and rang, but Frances had slipped away.

End.

The Gondolier, the ship on that young Frances waved to, was one on the MacBrayne steamers. At one time, they conducted a roaring trade, carrying passengers, from Glasgow or, Fort William to Inverness, via the Caledonian Canal and Loch Ness.

Six months after the outbreak of World War 2, the steam ship Gondolier went to her final resting place, in Scapa Flow, Orkney. Here she played her passive role as a sunken block ship closing one of the many narrow channels that led into the protected naval base.

Saying. I remember from my childhood.

The Lord gave man the earth and MacBraynes the western Highlands.
(Anon)

MacBrayne was a big business that ran the busses, the ferries, and even our school bus.

One of MacBrayne's steamers.

Scapa Flow.

Is a large natural harbour on the northern side of the Pentland Firth that separates mainland Scotland from the islands of Orkney Islands.

Scapa Flow's use as a safe harbour, dates back to before the Vikings, but it was during the two World Wars that it became a major naval base. There is a string of islands on the eastern side of the bay; during the First World War, the navy sank 20 ships to close the gaps. In 1916, the British fleet set out from here to engage the German fleet in the Battle of Jutland, (31 may 1916.)

It was here at the end of that war, that the Admiral of the 74 interned German fleet, ordered his captains to scuttle their ships. 30 ships were salvaged for scrap in the 1920's and 30's.'

One month after the outbreak of the Second World War, a German U-boat, U 47, found its way into Scapa Flow, between a block ship and one of the islands. It sunk the Royal Oak with the loss of 833 lives.

More block ships were sunk, (among them the Gondolier,) to fill in the gaps. Work then begun on the, 'Churchill Barriers' the causeways we see today. During the Second World War, 30,000 service men and women lived and worked at Scapa Flow. The first British civilian, of the Second World War died here during a bombing raid.

But the above is just history until you walk around the lonely and wind swept war cemetery that overlooks Scapa Flow, and read the words on the gravestones. Then you realise that beneath your feet, lay the bones of 14 and 15 year old boys, with titles like, 'stoker fourth class.' Then you understand that it was real people like you and I, who fought and died horrible deaths.

This is without doubt one of the most moving places I have ever visited. Close by there is a small museum inside an old oil storage tank, so if you are ever lucky enough to visit Orkney it is well worth a visit. <www.cantickhead.com/museum.htm>.

You will find a visitor book at the war cemetery; look for Yacht Scotia, my last visit was in June 2001. An Italian prisoner of war camp, occupied one of the islands at Scapa Flow. The church they built is now a minor tourist attraction.

An old custom.
In the town of Kirkwall, Orkney, they play a, New Year Bad' (Ball) Game.

The –game, is played between the Uppies and the Doonies, or (Up-the-Gates)' and (Doon-the-Gates) from Old Norse 'gata' (path or road). Today they use a ball, but this was not always the case. It is a wild and exciting game with few, if any rules, and dates back in time to the Vikings.

New Year
The Celtic celebrated it on, 'Samhain.' (November 1st).

Hogmanay.
(New Years Eve) is still a more important festival in Scotland than Christmas, and dates back to the Celts and the Norse, and may relate to, 'Hoggunott,' or 'Night of Slaughter'. The night the animals were killed for the midwinter feast.

We in Scotland have a rich heritage associated with this event, and we always try to get together with our family and friends to, see-in the New Year.

Now this next recipe is almost as old as the Scottish hills.
Black bun is very easy to make. We always have it on Hogmanay, but best to make it a few weeks in advance, and allow it to mature. Like all the recipes in this book there are many variants, so feel free to alter recipe to suit the ingredients available.

Black Bun
400g seedless raisins.
400g currants.
400g chopped, blanched almonds.
150g chopped mixed peel.
150g plain flour.
75g soft brown sugar.
One level teaspoon ground allspice.
Half level teaspoon of:-
Ground ginger, ground cinnamon and baking powder.

Pinch of black pepper.
One-tablespoon whisky.
One large, beaten egg.
Milk to moisten.
A pack of frozen short pastry (or make your own) to line the baking tin, keep back sufficient for a lid.

Mix the raisins, currants, almonds, peel and sugar together. Sift in the flour, the spices and baking powder. Bind them together using the whisky and almost all the egg and add enough milk to moisten. Pack the filling into the pastry lined tin, and add the pastry lid, pinch the edges together using milk or egg to seal really well. Lightly prick the surface with a fork and make four holes through the cake to the bottom of the tin with a skewer. Depress the centre slightly, (it will rise as it cooks.) Brush the top with milk or the rest of the egg to create a glaze. Bake in a pre-heated oven at 160°c for 2½ to 3 hours. Test with a knitting needle in the usual way. When cool store in an airtight container until Hogmanay.

The first of January.
When the New Year is only seconds old, we go, First Footing. The 'first foot,' to step into a house after midnight should 'ideally' be male, dark, and handsome, (well two out three's not bad. I have blond hair, well blondish. OK it's now a wee bit grey.) The first Foot should carry symbolic coal, shortbread, salt, black bun and, of course, whisky, and yell Happy New Year.

One for the Pot.

Poaching has become a dirty word of late. When we hear the word mentioned, we automatically imagine some native or other destroying the wildlife of his or her country by killing endangered species for their ivory, horn or even the bones of their victim's bodies.

In Scotland, not that long ago, a bird from the forest, a deer from the hill, or a salmon from the river, was the right of every man. Perhaps not a legal right, but a right never the less, with absentee landowners turning a blind eye. Well what else could they do staying so far away?

This story is about a poacher. Not the bloodthirsty villain the powers-that-be talk about, but an ordinary man. A man who takes, 'one for the pot,' now and then. Our poacher is a born and bred Highlander, a man with a family to feed. A man who works on the land his ancestors fought and died for. A land owned by the said absentee landowner. This man, this working man, whose job description, is farm labourer. Not a very apt description for a man who reaps and sows. Drives and repairs big tractors or combine harvesters, looks after herds of cattle or flocks of sheep. A man who is out in all weather delivering lambs. Och well I'm sure you get the picture.

Why, you ask does he need to poach. Well strictly speaking he does not, but like you and me he likes a bit of fun, a bit of excitement, a wee thrill and the extra food he brings in is always most welcome. He can't very well go out and shoot the pheasants that the estate keeper rears; they are reserved for the owner and for his or her rich guests. He might just pick up one or two, if they accidentally got themselves tangled in a wee bit of cone shaped wire netting, he carelessly left lying around among a bit of spilt grain. The absentee landowner, 'or Laird as he's known,' couldn't possibly mind if he picked up a bird that just happened to have a broken wing. Poor thing would be suffering, and easy prey for the vixen. The Laird would have no use for the hapless bird anyway. If it couldn't fly, he couldn't blast it out of the sky with number seven shot. Anyway, this day I'm telling you about, had been a hard one, it was just one thing after another. First, the tractor got a puncture and he had to fix it. That unexpected job took most of the morning. Then a tourist, who was gawking at the scenery instead of the way ahead, ran off the road, knocked down a fence killing a yew in the process. So he had to bury one and fix the other. Just in case you're wondering, it was the sheep he buried.

When he finally got home to his tied cottage, it had just stopped raining and the sky was beginning to clear. It had been a cold driech kind of afternoon, wet and miserable. The kind of day you and I would rather be inside with a wee dram and a good book. But never mind, he was home now, and he had the day off tomorrow.

As he opened his cottage door, he heard the baby crying and the other kids noisily squabbling over the TV's remote control. He entered his home and surveyed all he did not own. He removed his muddy boots at the door and carried them into the kitchen /sitting room. As he placed them on the newspaper by the Aga, he smiled at his wife who was standing behind the ironing board.

'Hi flower,' he said, but she just glowered at his wet boots as they leaked muddy water on the highly polished plastic tiled floor. No words of warmth for him, no have you had a busy day darling? Or let me pour you a wee toddy. She just barked.

'See if you can get the kids to shut up'. The kids succumbed to their father's fiercest look, all but the baby that is. Her proud father placed his big calloused hands into her tiny cot and ever so gently lifted out the newest addition to the family. He lovingly held her before him and felt her softness and smelt her powdery baby smell. She was just perfect, and as a smile spread over his weather-beaten face, the bad day began to fade from his memory. He held her up and blew a gentle raspberry into her little round belly, feeling the soft smooth skin on his nose. She spewed up her milky puke all over his hair and face. His wife screamed at him as if it was his fault.

After a quick bath, in lukewarm water, he sat down on his chair by the warm stove. His meal that night was a plate of potato soup and a 'mealy pudding' with mashed 'neaps and tatties,' followed by a mug of tea and a jammy piece. When the kids went to bed and the noise died down, his wife came and sat by his side. She told him, none to gently, that their car had failed its MOT test today, and that the garage wanted more than it was worth to fix it. He sat by her side, ran his fingers through her messy hair and promised to spend his day off, trying to patch it up.

He sat cuddling the wife that he loved, and thought about the night ahead. He wanted some proper food in the freezer and he knew just where to get it. Three weeks ago, his wife had cooked the last of the meat, and he had not tasted a salmon for months.

The trouble was the new farm manager, was a college boy, and a townie at that. The boy thought he knew it all, and took every opportunity to spout forth his book-learned knowledge. He was also opposed to killing the furry little animals that roamed the farm eating the crops.

Our farm worker stopped carrying his .22 rifle in the tractor under threat of the police and his job; so no more rabbits. He had even lifted his snares and had not been out hunting, or poaching, as the other people call it, for over three months. Not since the old manager retired. But tonight he was going out, the estate keeper had told him with a wink, that he was going away for the weekend, to a wedding in Inverness and that the new manager had gone away back to England for a few days holiday.

He opened a bottle of whisky and had a wee dram to keep the cold out. Of course, he knows that whisky doesn't keep out the cold. It just dulls the senses a bit so you don't feel it. Anyway, after his dram he put on his damp boots and set out. He headed in the general direction of the wood that lay half a mile behind his cottage. He sat down on a log and watched as the sun went down behind the mountain. Tonight was to be a full moon, the sky had cleared nicely and it was going to be a fine night. He continued to sit still for some time just watching, listening and enjoying the solitude. When he was satisfied no one was about he slipped off the log and went into the wood. He quickly located his hiding place and retrieved the plastic pipe that he had scrounged from a jobbing plumber. He unscrewed the end cap and withdrew his old rifle, then carefully un-wrapped it, from its protective oily rag that kept the rust at bay. Along with the rifle, he kept his dwindling supply of ammunition; he had been unable to get any for years. Not since the London government changed the rules. He knew he had no chance of ever getting a firearm certificate, that would require access to land to shoot on and he could never afford the cost.

His unlicensed .22 was an old bolt-action Winchester. It had belonged to his late father, who years ago had made and fitted a silencer to deaden the report. All that could be heard when it was fired was a wee phut. Assuming he used sub-sonic ammunition that is. From his kneeling position, he checked the wind direction and looked up at the rising moon, then slowly rose. He was now in hunting mode, alert and watchful. Barely making a sound, he crept deeper into the dark wood. Each foot as it reached the ground made a sideways motion to remove any twigs that might just snap under his weight. The night breeze was light and he was aware of every sound; more to the point he knew what made the noise. When he heard a rustle in the undergrowth his minds eye showed him the picture of a wee pointed nosed shrew, or a field mouse, and as he silently walked on, he heard the grunt of a hedgehog and the far away hoot of a barn owl.

It took him half an hour to reach the far side of the wood. He stood looking over a field dappled in opaque light. As he carefully scanned the drills of neaps, he saw a pair of brown hares boxing in a shaft of moonlight. For some reason he did not understand, hares had become scarce over the last few years. That was the reason he would never dream of shooting one.

Beyond them a dozen rabbits greedily nibbled on tender green shoots. The half-grown turnips with their tempting leaves and small purple hand size fruits were an irresistible lure to the rabbits and roe deer alike. This field had supplied the meat for his table for the last ten years. Over the last few weeks, he'd seen a lot of deer tracks. He knew the roe would visit this night. Trees surrounded the field on three sides, and cast deep shadows around the margins. He knew every inch of this remote patch of land and knew the forest beyond. He also knew the exact place where the deer would emerge. He settled down to wait with his bolt-action rifle in his hands, a round already locked in the old chamber. As he waited, he was aware of the moon as it floated higher into the night sky. He watched as its light rounded up the shadows, then herded them back to the very edge of the forest. An hour after he arrived, a subtle movement caught his eye; the first roe of the night had arrived at the far edge of the field. A long moment later, it took another step and the hunter saw a nice big buck with a perfect set of gold medal antlers.

Our poacher had no intention of shooting this proud buck. He knew the gamekeeper would be well aware of its existence and probably had a client lined up on the side, someone who would more than likely pay him good money for the rare trophy. What our man wanted was a doe, not too old, just fully-grown. The venison would be just sublime, just perfect.

Within half an hour, eight more roe deer had joined the big buck and slowly spread out over the field. Our hidden hunter saw the doe he wanted twenty minutes before. She was slowly moving in his direction, a nibble here and a nibble there, stopping occasionally to chew or look around. The owl hooted above his head and all the deer looked up and stared in his direction, then one by one, they dropped their heads and continued feeding. The light wind died away and our hunter was worried in case she scented him. But she did not; she just moved a few steps closer to nibble on another purple fruit. At last, she was so close the hunter could smell her. She was less than twenty feet away when the single, almost inaudible shot, found its mark and the doe silently dropped dead. The faint sound made the other deer lift their heads and look around the field, but seeing no movement, their heads soon dropped again. The hunter removed a white hanky from his pocket and waved it once in the air. The buck caught the movement and silently ran from the field followed by the rest of the timid herd. A few minutes later, from deep within the forest, the buck barked a belated warning.

The remaining doe lay still in the drills of neaps, but our man never moved. He stood and listened for a few minutes scanning the night for movement before venturing slowly out into the empty field. He gathered the roe up, and threw her warm body over his shoulder, then carried her deep into the wood, to a spot he knew well. He slit her throat and pumped her belly with his boot to help drain out the warm blood. He split the skin behind her Achilles tendons, pushed a stick through both hocks and hung his prize from the branch of a tree. He carefully skinned her, laying the skin fur side down below the carcass. He gralloched (gutted) her, carefully allowing the viscera to slip out to land on the skin. He split her sternum then reached into the cavity and cut out the diaphragm and then the lungs. His last job was to remove the head and forefeet.

He wrapped the evidence in the skin, and stuffed the lot down an old badgers den and rolled a stone over the top. He headed for home carrying his prize over his shoulder. On his way, he returned the rifle to its hiding place, and as was his custom, he continued to move slowly and silently. When he saw his cottage with one kitchen curtain open, and a faint light from the room beyond he relaxed. This was the all clear. No unexpected visitors and no surprises. He entered the kitchen and dropped the venison onto the wooden kitchen table, washed his hands and poured himself a much-deserved dram. Within half an hour, he had reduced the carcass to roasts, chops and stew. He gathered the skirt (thin meet that covers the stomach) and the scraps that remained and fed them into the hand-cranked mincer. He put the liver, heart and kidneys into the fridge and placed the meat on the larder shelf to hang. All that remained unused was the hind feet, so he wrapped them up in newspaper and put them into the back of the fire in the Aga. He had another wee dram, cleared up, and then headed for his bed.

This was not the only way to take a deer, but it was his favourite, quiet and clean. In time gone by the snare was widely used and perhaps still is in some places. He himself, as a boy used deer snares, trouble is you had to leave them over night, and if someone stumbled across it, they could sit and wait for your return. Besides, setting the snare took care, and care took time. The secret of a good poacher was to come and go unnoticed, and leave no trace of your visit behind.

The next morning he wandered over to the wood, just to make sure there was no sign of last night's activities and to walk passed his rifles hiding place. Just to be sure he'd hidden it well. He had one more night when he could be sure, or fairly sure, his nocturnal activities would go undetected. The river and the prospect of a few salmon in the freezer beckoned him. The taste of fresh wild salmon was very tempting. But he knew the water bailiffs watched the river. These men, employed by the Fishery Board, and whose wages were paid by a levy imposed on the riparian owners, would know the keeper was away, and would more than likely be about tonight. So our man headed for the sea. The seashore was not one of his favourite haunts; the sounds of the sea could deafen his ears to danger, but it was a lot safer than the river.

The tides were high just now. Spring tides the sailors call them. He arrived just on dusk and saw the tide was out; as he knew, it would be. He sat himself down in the cover above the beach to watch. It would be a long night. When the darkness finally fell, he wandered down to the beach with his long monofilament net in a sack over his shoulder. Keeping a wary eye on the sea for the bailiffs patrol boat, he searched for and found a brick size stone, he tied the end of the net to it and buried it in the sand, just about two inches down. Then he wandered backwards up the beach at an angle towards the river mouth allowing the net to flow out of its bag. At the high water mark, the net ran out. He uncoiled the rope attached to the float line and tied it to a tree. He wandered back down towards the sea spreading the net out flat, then returned to his hiding place and remained perfectly still. An hour later, when the moon came up, he could faintly see his net lying on the sand. He walked along the shore, away from his net to his waiting place. Not too close, but close enough to see the net, and more important to see a stretch of beach, in both directions. As the night wore on, he dozed. He hardly slept atall, his senses tuned to every faint familiar noise of the night. As he rested, the tide slowly crept in and covered his net. Now and then, his eyes would open and without moving, he would check the progress of the tide.

Now four hours later he could just see the cork floats strung out in the moonlight over the glassy water. He slowly and silently moved from his hiding place and crept along the edge of the wood. He stopped and listened again, then went down to the waters edge. He held the float line in his hand and smiled as he felt the tell tail jerking and tugging of an ensnared salmon. Glancing at the luminous hands of his watch, he saw that it was three thirty. A second fish suddenly became entangled in his net and over the next few minutes he felt two more and then another. That will do, he told him self. Greed, his father had taught him, was many a mans downfall; just you take what you can use. He heaved on the sole rope. When it came free, he hauled in the net, removing the fish as he came to them.

He landed six salmon in all, a good catch; he quickly retreated to the cover of the wood to pack his net and salmon into his bag. Now and again, he stopped his task to listen. When he finished packed, he stood up quietly and listened again. He had heard nothing, but never the less he sensed that something was amiss. He kneeled down again, and as quietly as he could he pushed the evidence under a bush and crawled away. He stopped every two yards or so to listen, but only heard the sounds of the night. He began to doubt his instincts, so he sat down and strained his ears, listened to, identified, then dismissed every sound he heard.

He had travelled perhaps thirty feet, when suddenly he heard a twig snap. He froze again as he heard the faint sound of a footfall. Someone was out there moving towards the shore. Slowly he lay down, he could now hear them breathing, they were only a few yards away and he knew he had crawled in the wrong direction. For twenty minutes, he remained perfectly still, hardly daring to move. Suddenly he heard the crackle of a radio, and a voice whispering.

'No it's all quiet Stuart; do you fancy a coffee? OK, I'll meet you back at the van.' With that, the water bailiff appeared and our intrepid poacher saw his outline against the sky as he passed by, only inches away. Then he was gone, silently back into the darkness of the wood. That was too close for comfort, if he had remained ten more minutes on the beach; he would have been caught red handed. He would have lost his job and been thrown out of his tied cottage. Then the local JP, who just happens to be a landowner, would have imposed a hefty fine. That was the last time he poached for salmon, from now on he would stick to easier prey.

End

A Gaelic proverb. *A switch from the forest, a bird from the hill, or a fish from the river were the natural right of every Highland gentleman.*

'**A switch**,' my interpretation of the word switch, is a stag with few or no branches on its antlers, but I think this proverb refers to a flexible branch cut from a tree used for making a bow. Or perhaps it just means firewood.

Highland Gentleman. A polite and generous man. A man who bought you a drink. The man, who in The Grouse and Claret, gave Dan the trout fly.

Lovat Scouts
The Highlander, and in particular the keepers, water bailiffs, stalkers and shepherds, are in tune with their environment. And like our poacher, spent their days observing, seeing and if required, not being seen. In 1899, Lord Lovat raised two companies from

among these men for service in the South Africa Boer Wars. They were highly successful as scouts and intelligence gatherers. In 1916 they were amalgamated to form the 10th (Lovat Scouts) Battalion Queen's Own Cameron Highlanders.

My grandfather served in the Lovat Scouts. And my father served in the Cameron Highlanders.

As I mentioned, Mealy Pudding in the yarn, I thought you might like to know how to make this ancient Scottish dish.

Mealy Pudding
Half a pound of oatmeal.
Two finely chopped onions.
One teaspoon salt.
If you like pepper, add about a half teaspoon of the stuff.
Half a pound of fat, suet or lard.

Ouch, this traditional Pudding was definitely not on your diet. Remember you can substitute oil, about 75 percent by volume.

Mix the lot together then dip a cloth into boiling water, (cheesecloth is ideal.) Take it out and rub with flour. Put the mixture into the cloth and loosely tie, or sew up. Place in a pan with a little water, about a third of the way up the pudding and boil for three hours, topping up the water as it boils away. Serve as-is, first day and when cold slice and fry. Goes well with neep's and tatties.

This is another ancient Highland recipe. Biscuits (cookies).

Sweet oat biscuits
1 breakfast cup of Rolled Oats.
75g Sugar.
1 teaspoon Baking Powder.
25g Butter.
1 Egg White.

Melt the butter and mix it with the oats, adding the baking powder and sugar. Beat the white of the egg until stiff, and stir it in. With floured fingers, form the paste into marble-sized balls. Place on a buttered tray and bake for 20 minutes in a slow oven (180°c). Nice, kinda crisp on the outside and soft in the centre. But it's a bit on the sweet side for me. Perhaps less sugar will work fine.

Sweet oat biscuits 2.
150g butter.
1 tbsp honey or syrup.
100g sugar.
200g oats.
Pinch salt.

Melt the butter in a pan, add the syrup and sugar and then mix in the salt and oats. Place mixture onto a greased baking tin. Pack down to a thickness of about ½". Bake at 180°c for about 30 minutes. When hot, score into squares.

Tattie soup.
Or Leek and Potato Soup, or whatever you have handy, I always include onion, turnip and carrot. This is a simple recipe for a tasty, hearty soup, popular throughout the Highlands. If you happen to have a bone so much the better.

Leek and Potato Soup,
Six tatties, diced.
1 leek, chopped.
1 onion chopped small.
1 small turnip chopped.
1 carrot, sliced.
Chicken stock, or boil up a bone.

Boil the potatoes, leeks, and other vegetables until soft, add seasoning.

Cock-a-leekie soup.
1 old hen or cockerel.
1 kg of leeks cut into 1-inch pieces.
Stock made from a marrowbone.
Salt and pepper to taste.

Believe me, this recipe was the only way to use a tough old

Put the lot in a pan, cover, and simmer for 4 hours. Remove from the heat then skim the fat from the surface. When cool remove the now hopefully edible cock or hen. Remove the bones, cut the meat into bite size pieces, return meat to the pan, heat up and serve.

Some cooks add a few stoned prunes to this soup, but my mum never did, nor do I. But if you want, you can add them with the chopped meat and bring to the boil.

The Huntress.

She was a wary creature. Some would say sleekit, (sly.) Certainly, she was canny. (Cautious.) She would never walk across an open field in daylight. Even at night, she preferred the shadowy margins, or stayed in deep cover. Out on the open moor she would follow the gullies between the peat-hags, or further out on the open hill where she felt a bit safer, she would patrol the heather moor looking for eggs or mice. If she was very lucky, she would catch a slow to rise nesting bird or even find carrion.

Last night her mate failed to return from his hunting trip. She greatly feared for his safety, for she had seen the bright light, and smelt the stink of man on the wind. Then just before dawn, she heard the loud noise that shouts death. She shivered as the deadly crack echoed from the glen. Instinct told her that she would never see her mate again.

The cubs within her bulging belly kicked and moved, reminding her to hurry up and finish her preparations. She had already cleared out three dens and was searching for the last, and perhaps the most important. This one had to be just right; she would only use it for a day or two, no more. Then she would carry her cubs, one at a time to another den.

She carefully searched down the banks of a burn looking for this den; the den she instinctively knew must be next to water. When at last, she saw what she was looking for, she held her head high and sniffed the surrounding air; searching for the scent of man, or another of her kind. When she detected no danger, she slowly and hesitantly approached the dark mouth. Once again, she paused but this time and stretched her head forward until her nose just entered the den and carefully sniffed the dank air. When she was satisfied it was safe, she entered and set about her housework. Just a little light dusting and a sweep out, a wee scrape here and there.

She felt her cubs kicking again as they moved down a little further. She knew it would not be long now. She carefully left the birth den and went hunting. She was lucky for almost right away she found a big plump heron, probably blown inland by the last storm, and crash landed in the long heather. She pounced, then held it's thrashing body in a death grip. When it was still, she dragged her prize a little closer to the birth den and carefully pushed it into an area of soft peat bog with her snout. When she was satisfied it could not be seen; she left her larder to continue her hunt. Suddenly she heard a sound and froze. She stood still cocking her head, turning it this way and that, trying to pinpoint the source of the faint noise. When she pounced, not one, but two mice darted away into the heather, she pounced again and caught them both under her fore paws. She felt so good she took one between her teeth and tossed it high in the air, and watched the wriggling doomed creature closely as it reached its zenith, then gravity took hold of the helpless prey and it fell into her open mouth. Over the next hour, she caught six more mice. Then the cubs within forced her to return to the den. She curled up and went to sleep, deep within the safe dark chamber.

For several hours, she slept; and in the darkness, she dreamed of her mouse, and as she dreamt, her feet re-enacted the moment of capture. When she awoke, it was to give birth to seven kits; she lovingly licked dry her tiny wriggling offspring, suckled them and kept them warm throughout the cold night and into the next day. She ventured out for a drink only once, and on the second night, she moved her young family to a new den. She carried them, one at a time, taking a different route each time.

Over the weeks that followed, she moved home often, and as the cubs became stronger and hungrier; she worked hard hunting, doing the job of two. On the poor hunting days, she retrieved her cached food from her wet peaty larder. Then came a time of plenty. Red deer were giving birth all over the mountain and leaving tasty leftovers for her to find.

Her cubs grew fast and if she could count, she would have realised that her family had shrunk to four during the harder times. The strong and healthy cubs; now three-quarter grown, could no longer be trusted to stay quiet. Often they followed their mother around spoiling the hunt in their eagerness. Through out that long summer, she continued to provide for them, relying more and more on finding bugs and worms rather than hunting for warm food. She taught them by example, sometimes she would fail to kill her prey and allow the cubs to finish the job. As the cold weather of winter approached and the mountain peaks donned their white caps, she and her family slowly moved down the mountain towards the forest. She found a winter den for her family under the roots of a great fallen pine tree.

The winter was hard, and food became even harder to find. They hunted along the low-lying fields looking for dead or dying sheep. They hunted the riverbank, and fed on the exhausted spent salmon. The long nights of winter crept past, and by February, they were only three. Her two cubs, now full grown, never left her side as might normally be the case, they had learned to hunt as a team and they were now deadly efficient. Their new prey was the newborn lambs, often weak and shivering in the cold white snow. Our team of hunters slowly wandered along the field next to the wood looking for any opportunity. Twins were always a good bet, triplets even better. One or two of our huntresses would circle and distract the mother sheep while the third would strike from behind her back. The ewe was no easy pushover. They had to take great care to avoid her head buts, but it was three to one, no contest. They feasted well, sometimes killing just because they could; perhaps only chewing the ears of the dead lamb before getting bored and moving on. It was the hunt and the kill that mattered.

Their nocturnal activities did not go unnoticed. For that night, she saw the spotlights bright beam as it stabbed the darkness along the field and forest fence line. She feared the light and shepherded her family away from the danger, but the cubs were young, and like all young, they did not know they lacked their mothers hard won knowledge.

Then a few nights later, the strong willed cubs, buoyed up with the confidence of youth, slipped out on their own. When they saw the spotlight they almost fled, but curiosity brought them closer. Then they heard the cry of a wounded animal, perhaps it was a young hare and they walked forward listening for the sound. The dazzling bright light seemed to be as harmless as the moon, and their confidence grew, then the sound of a hurt animal came again and they eagerly ran forward.

The vixen heard the sounds of death echo across the hills, and she knew she was alone again. She wearily sat down on her haunches and hung her head low. Half an hour crept past then she caught a movement in her peripheral vision and sprung to her feet. A big red fox sat ten metres away with his head tilted to one side looking at her.

End

It could be argued, that foxes are like man; both hunt for 'sport.' I also know that nowadays the fox receives a great deal of sympathy from the general public.

I like most people, love to see a fox in the wild, as I love to see all wild creatures in their natural environment. But I can also sympathise with the shepherd, and can see things from his or her point of view.

One night a few years ago, I drove around the fields with a gamekeeper. In the distance, we saw a vixen and her two grown up cubs hunting lambs. Although the keepers' job was to protect the lambs, we could not but admire the skill and wisdom of the hunters, for as soon as the keeper turned on his spotlight, they disappeared back into the cover of the wood.

In the morning, I walked along the same field helping the distraught shepherd pick up, and bury dead lambs. Killed by the vixen and her cubs. Most only had the tips of their ears chewed.

Strange sound
The gamekeeper made the strange sound by placing his moist lips on the back of his hand, and then gently sucking. Go on have a go. With a little practice, you can produce a sound similar to that made by a leveret (a young hare) in distress.

Wild Cats. (Felis silvestris silvestris.)
Not domestic cats that have gone feral, but truly wild animals. They enter this world with their eyes wide open, claws drawn and spitting. Sadly, this creature has now become scarce, even endangered.

I remember my father took a wild kitten home in a bag. 'Here kids' he called, 'I have a wee pet for you.' We all eagerly gathered in the kitchen. He closed the door, then at arms length, up-ended the sack. Well this tiny spitting tiger hit the floor, leaped onto the wall and circled the room twice at head height, then dived under a cupboard. I fell to my knees to look at the kitten and almost lost an eye.

My mother no doubt thinking that a wild cat was perhaps not an ideal pet for her weans, opened the door. This was only a wee kitten, a day or two old. A full grown wild cat will attack a man if cornered.

It is said that it is impossible to tame a wild cat kitten.

Clapshot; bet that draws a blank look.

This is a simple traditional recipe from the Orkneys. We had another name for it as kids; we called it leftovers. In the Highlands we call 'swedes' turnips so when I say turnip I mean swedes.

Recipe. Clapshot;
400g boiled potatoes.
400g boiled turnip.
1 or 2 tablespoons chopped chives, or whatever you have handy.
Salt and pepper to taste.
50g butter, margarine or oil.

Beat the two vegetables together while still hot and mix in the butter, chives and seasoning. (Or any thing else that you happen to have handy.) Beat in the pot until it is piping hot before serving. Also very good if made into cakes. Dip into egg, cover with breadcrumbs and fry.

Minced Collops. Oops, another puzzled look.
400g of minced steak.
50g of butter, or oil.
3 small onions.
½ a pint beef stock or water.
Salt and pepper to taste.

Fry the onions in a little oil for a minute or two then add them to the mince. Brown, make sure you remove all the lumps. Add stock, or water, season. Simmer for 1 hour; remove from heat and skim of the fat from the surface. Add a handful of oatmeal. And cook for a few more minutes then serve with mashed tatties.

The Wee Lairdie.

When Jack MacTavish reached the age of fourteen his childhood and schooling came to an abrupt end. He assumed he would work alongside his father and learn his trade. His father, who was a joiner on a remote Highland estate, was a little less confident. He knew that ever since the end of the World War, the country was in what the newspapers called, a deep recession. He had read that thousands of men were without work. Although there was little sign of the recession on the estate, he knew from the odd newspaper that came his way that times were hard.

The joiner had mentioned his son Jack in passing to the Factor, and told him that the boy would be leaving the school soon; he even hinted that he was in need of an apprentice. The Factor appeared to take an interest, and said he would discuss the matter of an apprenticeship with the Laird. That was over a month ago and the delay and worry had once again brought the boys father to the Factors office.

The joiner, dressed in his brown bib and braces dungarees, carrying his flat cap in his hand, knocked on the office door.

'Come in,' a voice called from within.

'It's about my son Jack sir. It's today he leaves the school and I was just wondering if there was any word about a job for him.'

'Take a seat Willie,' The Factor said waving at a chair.

'Thank you sir, I'm fine just as I am.' Willie replied as he kneaded his cap in his nervous hands.

'I have mentioned your son to the Laird, on two occasions,' the Factor replied, 'and the Laird is considering your request.'

'Thank you very much sir,' Willie said wondering if he dare say any more.

'I have a meeting with the Laird today,' the Factor continued, 'and I'll mention your boy to him again. Now I can't promise, but I believe the Laird will offer your son the apprenticeship.'

Willie left the office and went back to work believing his request would now be granted. A few days later the Laird walked into the workshop, the joiner naturally assumed permission was about to be given. He was however surprised that the Laird had chosen to come in person to give him the news.

'Good morning Willie,' the Laird said in a loud voice. Willie doffed his cap and returned the customary greeting.

'It's about your son,' the Laird continued, 'the Factor asked me if we might find a place for him on the estate; well I am sure we can. Your boy by all accounts has done well in school, and I have heard very good reports about him. What I suggest,' the Laird continued, 'is that you send him up to the house tomorrow morning. I would like him to spend some time with my son Charles. They are both the same age and I believe they share an interest in fishing. When Charles goes back to school, we can decide what to do with your boy.' Willie looked a bit bewildered by this unexpected statement, and the Laird seeing the look on Willies face assumed that Willie was about to turn down the offer.

'I'll see to it that you are not out of pocket by this arrangement,' he generously added.

When the Laird asks, few would dare refuse, and an estate worker who lived in a tied house with his family would never dream of refusing any request from the big man, so he thanked the Laird and expressed his deep gratitude.

When he got home that night, he informed his son that the Laird wanted him up at the big house.

'What for?' his son asked, a little afraid of the thought of visiting the Laird in his home.

'It seems the wee Lairdie is lonely,' he said quietly, 'so you just away up there in the morning and do what they ask.'

The son of the big house was an only child. His mother had died a few years ago, and his father had decided that what his son needed was another boy for company, while he was home from his boarding school. The Lairds son Charles was a thin and delicate looking boy, and his father thought he was lonely, and young Jack was available. He hoped that some of Jacks energy and strength might rub off onto his son, and that his son might lose some of his softness.

As it turned out the boys hit it off from the start and rapidly became friends. They spent the long summer day's bird nesting, wandering the hills or fishing the hill lochans.

When Charles went back to school he looked a lot fitter and healthier and his father thought he had even developed a few muscles. He sent for Jack, they met in a smallish room, small for the big house that is; it was comfortable, with an open log fire, deep chairs and two sofas with soft arms you could sit on. The walls, lined with dark wood panelling, were stained with years of peat smoke, but best of all this room held a full size billiard table that the boys had well used over the summer.

'Master MacTavish,' the butler announced as he showed Jack in.

'Ah Jack, come away in,' the tweed clad man said, 'Take a seat,' he ordered waving his hand in the general direction of an armchair by the fire. 'Well Jack,' he continued, 'how have you enjoyed your summer with Charles, better than school I dare say?' The village boy sat on the very edge of his chair, a little in awe of the big man. 'Good sir,' he managed to say.

'Splendid,' the Laird replied, 'now what are we to do with you now that Charles has gone back to school. We will have to find something to keep you out of mischief. A job of some sort perhaps, what do you say to that?'

'Thank you sir,' young Jack answered respectfully.

'Well now, what do you want to do Jack, are you an inside man or an outdoor man?'

'Outdoor sir, I think,' Jack replied.

'Do you want to work with your father or would you rather work with the Factor. I have had a word with him and he is agreeable for you to work with him. Yes that's best I think, good Jack, I'm glad that's settled. You go and see the Factor.' With that, the Laird stood up and wandered out of the room, the meeting was apparently over. Jack stood by the fire not sure what to do next.

What would his father say; he was very keen on apprenticeships. A few minutes later, the man dressed in a dark suit returned.

'Well Jack, it seems the Laird likes you,' the butler informed him, 'I'm to take you over to the Factors office now, so if you're ready.'

Jack of course knew the butler; he had always been a generous man and had treated Jack and Charles as if they were brothers. Half an hour later, Jack sat in the Factor's office. The Factor was a thin man in his mid sixties dressed in tweeds, his hair was white, and he spoke just like the Laird; Jack reckoned it was a public school accent. The Factor asked Jack a lot a questions about his schooling, and then asked him to add up a column of figures. After checking Jacks sums and finding them correct, the Factor spent the next hour carefully explaining that he was really an agent. He managed the property side of the estate for the Laird, collected rents and was responsible for property maintenance.

'The Laird has asked me to employ you Jack; I'm to teach you my job, and as you know what the Laird wants the Laird gets,' he added with a smile, 'but seriously Jack, this is a good opportunity you have been given and I am happy to have you with me. If you work hard as I'm sure you will, your future will be secure.'

After the pep talk they went out in the buggy and the Factor asked Jack to drive. As they travelled around the estate the Factor pointed out this or that. The Factor handed Jack a notebook and told him to make notes about work required. They went out twice a week Tuesday and Thursday. Within two months, Jack had visited every farm and tenant on the estate. Jack spent hours reading old correspondence to and from tenants and he was never afraid to ask if he failed to understand something, and when he came across an error, his natural tact greatly endeared him to the Factor. The Factor was amazed how quickly Jack grasped the workings of the estate, and had no hesitation in giving him the responsibility of the big ledger. Now, when a rent was paid, Jack entered the amount into the book. If a tenant asked for repair work on their property Jack and the Factor would carry out an inspection, and then decide what action to take. If the tenants had other problems, or major work was required, the factor wrote out a report and sent it to the

Laird. Within six months, Jack had earned the factors trust, and given the authority to inspect, and authorise minor repairs.

When Charles came home, the Factor gave Jack as much time off as he wanted, or rather as much time off as Charles wanted.

Jack continued to learn under the Factors teaching and spent all his free time studying books on all aspects of estate management and silviculture. (Growing and cultivating trees.)

This arrangement continued throughout the rest of Charles schooling. When Charles left university with a law degree, his father was very proud and felt sure that putting him and Jack together had helped his son a lot.

Time passed then suddenly a second world war was thrust upon them. One became an officer the other his batman. The war was hard on Charles, although he became an officer by right of birth; he lacked what it took to lead men and spent most of his time in administration.

Jack on the other hand was a born leader, but when Charles arranged and offered him the chance of a commission, he turned it down. 'I'll just stay with you Charles,' he said, 'someone has to keep you right.'

When World War 2 reached its bloody conclusion, both men returned to the Highland estate. The old Factor was delighted to see Jack again and within a few months, he retired leaving Jack in charge. Within a year Jack married a girl from the village and they settled down in a cottage by the loch. The Laird grew old and Charles reluctantly took on his fathers responsibilities. Jack accompanied Charles to meetings and helped him as he learned to run the estate. When the old Laird died, Charles became Laird. He asked Jack to move his office into the big house, and Jacks job slowly changed. He became a sort of secretary for a time, then an intermediary, but more than that really. Charles had no interest in running the estate and disliked dealing with people, or making decisions, so he gradually pushed more and more of the work on to Jack. Jack just sort of took over, not in a pushy way but gradually. He became the man to see, a man who could make a decision quickly, he was always extremely fair and impartial and so Jack deservedly gained the respect of the workers and the tenants.

As time passed, Charles became a bit of a recluse. Sometimes he stayed in his room for days on end. During his times of solitude, he would only speak with Jack, or to the old butler. When the latter retired, he was not replaced. Time slipped past and the years wore on.

Charles and Jack still went fishing now and then; mostly it was Jack who persuaded Charles to take to the hills. Once out Charles always enjoyed himself, they would laugh and joke like old times. They also enjoyed a bit of pheasant shooting, not an organised shoot, just a quiet walk through the woods with gun and dog.

Time relentlessly marched by and they grew old together. Jack was blessed with a son and daughter. Charles, who never married, became godfather to them both. The children were bright and went through university. Jack wanted them to work on the estate, and his son now did the fathers job. His job description is estate manager; Jacks daughter preferred town life and became a solicitor in Inverness. Both were now married and Jack was a grandfather.

One day Charles asked. 'How old are you Jack?'

'You know the answer to that as well as I do,' Jack replied to his friend, 'but since it's you asking I'll tell you, I am 68 years old next April, same as you.'

'Well,' Charles said, 'it seems as if time has speeded up a bit, I think you should retire.'

'I am retired,' Jack replied, 'this place runs its self, we have good office staff and good workers and you and I are surplus to requirements.'

'Well, if that's the case let's do something.'

'What do you have in mind?' Jack enquired.

'I don't know,' Charles replied, 'just something before we get old.'

Jack the manager as always said, 'right, say no more now, just get a note book and write down every crazy or not so crazy idea you can think of. I'll do the same, tomorrow we will compare notes and make a decision.'

'Splendid,' Charles shouted. Jack went home happy that Charles was so cheerful; and that evening he sat by his fire and compiled his wish list. Next morning when Charles failed to appear. Jack went into his bedroom looking for him. He found that Charles had died in the night. By his chair sat his wish list. Jack picked it up and with the tears running down his face put it in his pocket.

Jack was lost; his lifelong friend was gone. When Jack eventually returned to his wife on that saddest of days, he sat by the fire, thinking about Charles. He remembered the wish list, and slowly retrieved it from his pocket and stared at it in disbelief. He picked up the list he had made and held them side by side. The lists were identical, almost word for word.

Several days later, Charles's solicitor arrived at the estate. Jack learned that apart from some small bequests, the entire estate came to him. Jack had become the new Laird.

Next day Jack did the first thing on their lists; he bought a small yacht and named it. *The Wee Lairdie*.

End

How to become a Laird or Lady.
The title Laird is associated with Scottish sporting estates and land ownership, but you too can become a Laird, or Lady.

How? You ask. Easy, go onto the internet, type in 'Laird.' There are lots of people out there selling a square foot of Scotland, along with the Laird title. But what's a title. It's just another name, and as such, it is meaningless. It certainly won't make you a better person. Robby Burns once wrote. 'A mans a man for all that.'

While first footing, we used to carry a little shortbread, a lump of coal and a bottle of whisky. Food, heat, drink and friendship. Shortbread is as old as Loch Ness and known the world over.

Shortbread.
150g plain flour.
50g caster sugar.
100g butter.

Sift the flour, add sugar, add the butter and knead until the mixture binds together. To make flat rounds, divide mixture into two, roll out to about ½" thick, pinch edges to form a pattern. Prick all over with a fork, score surface into 8 wedges. Place them in a lightly buttered baking tin and bake in a cool oven, 150°c for 40 to 60 minutes. Or you can form the mixture into fingers or fancy shapes.

Lots of recipes, but this basic recipe gives consistent good results.

Shortbread 2
150g Plain flour.
100g butter.
50g caster sugar.
25g corn flour.

Blend the butter and sugar together, using a wooden spoon. When creamy sieve in the flour and corn flour, mix well. Dust your board and the dough with flour and roll out to about a quarter inch thick. Prick over with a fork and cut scores on the top to form slices. Use your fork to make a pattern around the edge. Place the shortbread onto an oiled baking tray and bake for 25 to 30 minutes in a pre-heated oven at 170°c. When ready the short bread will be pale brown and crisp. Sprinkle with caster sugar, and place on a wire cooling tray. Store in an airtight tin once they are cold. Shortbread keeps well; if you hide the tin.

The Statement.

'Now Mr Mackay,' the man with grey curly hair and a deadpan expression on his face said, 'I want you to tell me about your day. Take your time. Start from when you got up this morning.'

'Well, Inspector,' the bank manager said as he leaned back in his chair and placed the tips of his fingers together. 'I got up as usual, nothing out of the ordinary. I remember I glanced out of my bedroom window at the snow. It was a bit windy and the snow had started to drift. I remembered wondering if I'd have to dig my car out. The only slightly unusual thing that happened,' he told the policeman, 'was the phone call I received. I was on my way out at the time, so I took the call on my hall phone. Some one asked if I was Mr Mackay, I said I was.'

'Man or woman?' the inspector interrupted.

'Oh! A man; he asked me if I was the manager of the High Street bank. I said I was and wondered how he got hold of my phone number. Anyway, when I asked him who he was, the line went dead. Then I drove to work.'

'Do you live alone Mr Mackay?'

'No, well yes… My wife is staying with her mother just now.'

'When do you expect her back?'

'I'm not sure, her mother is unwell.'

'May I have her address please?'

After the policeman wrote down the address, he looked up. 'Now Mr Mackay can you tell me why you arrived half an hour late at the bank this morning.'

'Was I?' he asked with a puzzled look on his face. 'Well I suppose I drove slower because of the snow'.

'Go on.'

'About eleven o'clock my secretary showed a man into my office. He had no appointment, but my secretary must have decided that I just had to see him.'

'Can you describe him?'

'He was a little taller than me, perhaps 1.8 or 9. His hair was grey with white streaks running through it, and it was sort of wavy. His eyes were a deep blue and... I'm sorry, that's all I seem to remember about him.' The policeman looked up then made a note in his book. 'Go on.'

'I do remember looking up from my paperwork,' he continued, 'and realized that my secretary never gave me his name. I thought that was odd and definitely not like her. Anyway as I stood up to greet my unexpected appointment my secretary left. I offered the stranger my hand and asked how I could help. As I looked at him, he held my gaze, and I was trapped, unable to move. I just stared into his eyes. I knew I was staring but I couldn't help it. Then I became aware of a weakness creeping through me and thought this can't be happening. I thought I was having a stroke, or some kind of fit.' He stopped talking and looked up at the police Inspectors deadpan face. 'My vision began to close in' he continued. 'All I remember is his eyes and a creeping emptiness flowing into my body. I just sort of drifted away as his eyes became a blur, then nothing.'

'Yes but that was,' the inspector said as he looked at his watch, 'six hours ago, is there nothing else you remember?'

'Well yes,' he said slowly, 'I remember clearly, but you won't believe me. In fact I don't believe it myself.'

'Tell me.'

'Alright, when I awoke the sun was hot, it was burning into me. I opened my eyes and slowly looked around and saw trees. To say I was confused would be a huge understatement. I sat up and saw my bare feet, they were very dirty and seemed bigger than they should be, certainly wider. My toenails needed cut and my trousers were in rags. I recognized them as my normal working trousers. I have three pairs the same at home. I lifted my hands and looked at the alien things before me, they were dirty with cuts, scabs and the nails were black and broken. I looked around the wood and tried to stand up, and found I could. I felt fit enough, in fact, I felt just as I always felt, except this was weird. I glanced at my watch but I had no watch. I felt for my wallet, but it too was missing, in fact, all my pockets were empty. I realised I was frightened, although what of, I had no idea.

I started to walk. Although my feet were bare, it felt as if I wore shoes. I walked and walked through open woodland but I saw no birds, no anything, just trees. I walked for about an hour before I came across a narrow track and wondered which way to go. Left seemed slightly down hill, so I went left. The track, I could see was only used occasionally, perhaps by a farm tractor or a four by four. I walked on and eventually came to a wooden gate; I climbed over and continued on my way. I walked on for about an hour, then came across a narrow road.

For no particular reason, I turned right and continued to walk. After what seemed like ages, I got very tired, so I sat down on a stone to rest. My fear returned but I decided it was just confusion. I knew who I was, and I remembered the visitor to my office. I remembered the strange feeling when he looked at me. I knew I must have had a fit of some kind. It seemed as if it was only a few minutes ago, but my state of dress and my beard. Oh! I had a beard, and as I rubbed it, I wondered how fast a beard grew. I tried to work it out and decided that mine grew about a millimetre a day, or seven mm a week. I came up with five months growth, and decided the answer was consistent with my state of dress. Five months was a long time but the evidence left no doubt. I began to think that I must have had a brainstorm or something. I remember closing my eyes for a second, and almost immediately, I was shivering with the cold.

I opened my eyes and saw frost on the ground and I looked at my feet and saw I wore a pair of black rubber Wellington boots. I instinctively rubbed my beard and discovered it was gone. I looked around but I was still in the same place, sitting on the now very cold rock. I stood and looked around. Now if I was confused before I was twice as confused now. I knew I was suffering from memory loss, with days weeks or even months missing. I started to walk again, but strangely, I was not hungry. I felt just fine. Only a bit cold. I checked my pockets for the umpteenth time. They were still empty. On an impulse, I picked a green leaf from a tree and put it in my pocket. I did not recognise the tree type, but then I wouldn't. I'm not a countryman; to me a tree is just a tree.

I followed the road for perhaps half an hour before it abruptly ended, well not ended but changed into a dirt path. I'd walked the wrong way, oh well! I remember thinking, there was nothing for it but to back track, so that's what I did. When I got to my rock, I was tempted to sit down again but resisted the impulse. Time! I remember wondering what time it was, and then realised that I was not cold any more; the sun was in fact quite warm. I passed by the rock and when I got to the locked gate, I passed that by with barely a glance. I was walking fast now, eager to reach civilisation but the narrow road seemed to go on forever. Eventually I came to a rock, it looked like the rock I had sat on earlier and with a sinking feeling, I realized it was the same rock. I stood still and stared at it. Suddenly I lost my energy, my shoulders slumped and I sat down. Then a second later, it was dark, but I had only just sat down. Or thought I had. I looked up and saw lots of stars that filled the black clear sky. I felt my beard; it was back but not as long as before, perhaps only a month's growth. I realised that my feet were once again bare. It was then that I decided I had gone mad. I closed my eyes to shield them from the bright sun that had suddenly burst out. I stood up, put my hand in my pocket and found the leaf. It was still green but I thought it had been a month since I picked it and knew it should have dried up and turned brown. Or at least withered, but it looked as if I had only just picked it.

I replaced the leaf in my pocket and looked around; I wanted to walk again, but which way. I remember trying to think logical thoughts and listened carefully, I knew I was in a wood, and I knew I should be hearing birds or animals, but it was silent. I pinched my skin, and said ouch aloud and decided everything was real. I started to walk again, slowly this time and without enthusiasm. As I walked, I examined the trees, but they were still just trees. I could never see very far ahead because the track twisted and meandered through the forest. I realised that I must have walked in a circle. I glanced up and saw the sun through the trees, then I decided that if I followed the sun, or rather keep it in the same place relative to me I should more or less walk in a straight line. I picked up my speed and I think my confidence returned a bit. I felt certain that now I would find my way out of the wood, but two hours later, my rock appeared before me again. I sat down confused and bewildered, and I can tell you I was very near to tears.

Then as I sat with my head in my hands, I heard a telephone ring. I opened my eyes and discovered I was back here in my office, sitting behind my desk. I looked at my hands and saw they were clean. My clothing was pristine and my shoes polished. My hand shot up to my chin and I felt smooth skin.

Just then, I heard a knock on my door and my secretary walked in. Excuse me Mr Mackay; she said in a voice that I thought lacked her usual confidence. I looked up and stared, her eye makeup had run down her face as if she had been crying. I asked her what was wrong. It's the safe sir, she sobbed out. Then she told me about the safety deposit boxes and that all the cash drawers were empty! I instinctively hit the emergency button with my knee. That's it Inspector, I know it's ridiculous but that's just as I remember my day.'

The Inspector looked up and slowly shook his head.

The bank manager put his hand in his pocket and removed a fresh green leaf.

End.

What has the last wee story got to do with Scotland you ask, well read on.

Hypnosis.

In the 19th century, 'Papyrus records, dating from the 3rd century,' were discovered at Thebes Egypt. In one of the records, we see the first written description of hypnosis.

'You take a boy and sit him upon another new brick, his face being turned to the lamp and you close his eyes and recite these things which are written above down into the boy's head, seven times.

You make him open his eyes. You say to him: 'Do you see the light?' When he says to you, 'I see the light in the flame of the lamp', you cry at that moment, saying 'Heoue' nine times. You ask him concerning everything that you wish.'

A few famous Scots.
Braid (1795-1860) Pioneered hypnosis. See that's the connection.

What other famous Scots are there? Well I'm glad you asked, but this book is not long enough to list them all so here are just a few.

William Paterson. (1658-1719.) Founder of the Bank of England. One of the businessmen who was an ardent supporter of the Treaty of Union. His motive, greed. See. Such a Parcel of Rogues in a Nation,

James Chalmers. (1782-1853.) invented the adhesive postage stamp.

Patrick Ferguson. (1744-1780) invented the breech loading rifle.

James Gregory. (1638-1675) invented the reflecting telescope.

Andrew Meikle. (1719-1811) invented the threshing machine.

John Napier. (1550-1617) developed the concept of logarithms and invented the decimal point.

Sir James Young Simpson. (1811-1870) pioneered the use of chloroform in anaesthetics.

James Small. (1730-1793) invented the iron plough.

William Symmington. (1763-1831) Developed the first steam powered marine engine.

Robert William Thomson. (1822-1873) invented the vulcanised rubber pneumatic tyre. He also patented the fountain pen, and the steam traction engine.

Sir Robert Alexander Watson-Watt. (1892-1973). Developed RADAR.

John Logie Baird. (1888-1946).Invented the television.

Sir Alexander Flemming. (1881-1955) Discovered penicillin.

Alexander Graham Bell. (1847-1922) invented the telephone.

James Young. (1811-1883) developed the process of refining oil.

Sir James Dewar. (1842-1923) invented the vacuum flask.

John Boyd Dunlop. (1840-1921) further developed the vulcanised rubber pneumatic tyre. (Dunlop Tyres).

Rev. Patrick Bell. (1800-1869) invented the reaping machine, which led to the combine harvester.

Sir Dugald Clerk. (1854-1932) invented the two-stroke Engine.

Rev. Alexander Forsyth. (1769-1848) Invented the percussion cap that later became the bullet.

John Shepherd-Barron. (1925-) Inventor of the ATM.

Sir Sandford Flemming. (1827-1915) created the World time zones.

Alexander Wood. (1817-1884) invented the hypodermic needle.

John J R MacLeod. (1876-1935) helped discoverer of insulin.

Henry Faulds. (1843-1930) created the process of criminal fingerprinting.

Ian Donald. (1910-1987) invented the ultrasound scanner.

One more, but first a wee question.

Who discovered, Natural Selection? You know the principle of evolution.

Was it Charles Darwin, Alfred Wallace or Patrick Matthew?

Answer. It was Scotsman **Patrick Matthew**, he published his theory in 1831 some 25 years before the fore mentioned. (Not famous but he should be.)

Recipe (not.)
Capercaillie is the largest of the highland grouse, (about the size of a goose) and is now along with the black cock protected. The black cock is the second largest grouse. So I'll give you an old wife's recipe.

Nail the bird to a stout board, dig a hole and bury board and bird, dig up six weeks later and cook the board. (Old Joke) The

Capercaillie feeds on pine needles and the meat tastes strongly of, well pine.

Fish soup, from the fishing village of Cullen, on the Moray Firth.

Cullen Skink
Smoked haddock, around 1kg.
1 medium onion, finely chopped.
1½ pints milk.
2 tablespoons butter.
200g mashed tatties.
Salt and pepper.
1 bay leaf.
Chopped parsley.
Water.

Cover the fish with water, bring to the boil and simmer for about 5 minutes. When cool remove the fish, remove skin and bones, flake the meat. Add the chopped onion, bay leaf, salt and pepper to the sk. Simmer for 15 minutes, and then remove the bay leaf. Add the milk to the fish stock, and bring back to the boil. Add enough mashed potato to create the consistency you prefer, (don't be afraid to make it rich and thick!). Add the fish and reheat. Check for seasoning. Just before serving, add the butter in small pieces so that it runs through the soup. Serve with chopped parsley sprinkled on top, accompanied by triangles of toast.

This is just one recipe for fish soup but you can make variants using prawns, salmon and other fish. In fact it's like all cooking, just take the basic recipe and do your own thing, be inventive.

When in France I tried their famous Bouillabaisse. It's OK but not a patch on Cullen Skink.

Recipe. Salmon Fish Cakes.
As kids, we were fed on a diet of salmon, venison and fresh vegetables from my father's garden. Plus heather honey and home made bread, cakes and jam. Did we appreciate this healthy diet, no of course we didn't. Fish cakes made from salmon was nice for a change, especially if we got baked beans with it. 400g of cold, cooked leftover salmon.

½ a pint of thick white sauce. (Optional)
600g cooked and mashed tatties.
Chopped parsley.
Salt to taste.
1 egg.
Breadcrumbs.

Mix together left over cooked fish and tatties. Add parsley and season; dip in beaten egg, breadcrumbs. Fry until golden brown. I seem to remember my mum some times served them with a poached egg on the side.

I'll sneak in one last bit of history.

The Stone of Destiny or, Stone of Scone. Ancient coronation Stone, dating from the times of the Picts. In the year 843, the first king of Scotland, Kenneth MacAlpine, brought the Coronation Stone to Scone. The crowning of Scottish kings continued on the Stone until 1296. On that date, Edward the 1st of England stole the Stone in a raid into Scotland and placed it under the English throne. The coronation of Scottish monarchs continued at Scone until 1651. The Stone of destiny was 'Liberated' in 1950 and returned to Scotland. A few months later, 'a stone,' went back to England.

Prime Minister John Major finally returned,' the stone' to the Scottish Parliament in 1996.

<u>Sandy's Widow.</u>

Sandy left me one cold winter's night in early December. I dashed from the house into the pouring rain wearing only my dressing gown and slippers and ran to the phone box at the end of the street. The ambulance soon arrived but I knew in my heart it was too late.

The ambulance men were very nice, but what could they say, my Sandy was gone.

We had been together for nearly thirty years, through the hard times and the good times, now I was alone.

The funeral was a quiet affair, some of the people from our street came and some of his old work mates volunteered to carry his coffin. We passed by the now closed ship yard where he had spent his working life building ships that sailed out from the Clyde and into the seven seas.

After the funeral, I sorted out his clothing and took them down to the hospice shop.

'Thank you hen. (Same as pal, love, dear etc) Just leave them by the door,' the busy lady said. I don't know what I expected, but I suppose they must get lots of stuff handed in. It's just that Sandy's things were special, now they too were gone.

That evening I took the suitcase from under our bed, the case Sandy kept all our bits of paper in. I slowly removed the documents and read each in turn. Sandy's birth certificate was on top. I then picked up our marriage certificate and smiled as I remembered our marriage, and the borrowed dress I wore. I thought about the registry office and the dance in the hotel. I remembered our shyness on our first night alone together. I cried when I found the photo of our only child who died when she was three years old. I forced myself to continue and picked up Sandy's union card, and remembered the long hard times with little money coming in when he was on strike. I found the letters I had written to him when he was in the army, and then his army record. I found a box containing four medals and wondered why he had never shown them to me.

I placed Sandy's death certificate in the case, and began to repack the contents. It was then that I noticed a small bulge in the lining. I felt the bump and found it to be hard. Eventually I worked the package loose and was surprised to discover it contained money.

Sandy, it seems, had secretly been saving up; I turned the envelope over and read the pencil words, printed by Sandy's firm hand.

FOR OUR OLD AGE AND A HOLIDAY FOR MARGARET.

I emptied the money out onto the bed and counted the notes, it was a fortune, nearly five hundred pounds. I replaced the money with shaking hands and sat on the bed thinking about the future. What was I to do with my life?

Now it's just me, so I'm free and I suppose I can do as I please, so what do I want to do?

Over the next few days, I thought about nothing else. Get out of Glasgow, kept coming into my mind, but where would I go, perhaps I could move back to the country, perhaps I could go back to the village I was born in and visit my parents, but would they speak to me. I was only sixteen when I ran away with Sandy and I had never dared to go back. That's it, I suddenly decided, I'll go back to my village for a holiday and visit my parents.

Next morning I walked to the council offices and used some of the money to pay the rent on our council flat. Then I packed a bag and took the subway to the bus station. I got the timing a bit wrong and had to wait two hours for my bus.

The journey took most of what was left of the day and I began to wonder if I had made a wise decision. I arrived in the late afternoon when it was just getting dark, and walked the last half mile to the village. As I walked, a light sleet began to drift down from the hills.

When I reached my parents house I stood shivering by the front gate, but my courage deserted me, so I walked on. I suppose I should have telephoned and booked in somewhere, but it never occurred to me. I looked around for a bed and breakfast, but saw no signs. I asked a man who was walking his dog if there was anywhere I could stay the night. He shook his head as he thought about my question.

'Yes,' he blurted out as he lifted his hand and touched his forehead in a mock salute, 'the Inn,' he said smiling, 'they have a room,' the smile faded and he added, 'well they used to have. Anyway you go and ask there.'

I went into the village pub and saw that it was very quiet, just the one workman sitting having a pint and chatting with the barman.

'Have you a room for let,' I asked. They both stared at me for a moment.

'You can come and stay with me,' the workman offered. I shook my head and took a step backward.

'Take no heed of him hen,' the barman said, 'you come away in out of the cold,' so I slowly walked up to the bar counter.

'Well now,' he said, 'we do have a wee room, but its no been used since the summer and it might be a bit on the cold side, but it's yours if you want it.' He called through an open doorway and his wife appeared a second later. She was dressed in jeans and wore a pink nylon housecoat. As she wiped her hands on a dishcloth, she said.

'Come away ben,' (into an inner room) we can have a nice cup of tea in the kitchen and you can get yourself warmed up, it must be freezing out there. Did you come on the bus?' As she led me into the warm kitchen I looked at her red hair, and something about her features brought back memories of my childhood.

'Was your name Elspeth Duncan,' I asked remembering a school friend, she lifted her head and looked at me.

'Heavens no, that was my mother's maiden name.' We sat in the kitchen and I told her about Sandy, and she told me about her mother who had died of cancer twenty years ago.

Eventually she took me up to a small wood lined attic room, with a steep sloping ceiling and we made up the bed and turned on an electric fire.

Next morning I wandered through the village lost in my past and slowly made my way to my parent's cottage. I stood at their garden gate again and looked at my old home. Eventually I gathered my courage and pushed open the gate, and walked up the path to the front door. Just as I raised my hand to knock a child opened the door and dashed out, he saw me and stopped and stared.

'Is Mr. and Mrs. Young in?' I asked.

'Who do you want?' the woman who appeared at the door carrying a shopping bag asked.

'Mr and Mrs Young,' I repeated.

'I think that's the folk who used to stay here,' she said, sorry I can't stop, I'm rushing to catch the bus.' With that, she dashed away leaving me to close the gate. Perhaps it's just as well, I thought my father would never have forgiven me for running away, and I remembered the blazing row when he found out I had a boyfriend. But it would have been nice just to see them again.

I wandered away thinking about my old home and remembering the happy times of my childhood; I looked at the small grassy play field and remembered the games of kick the cannie, (tin-can)and hide and seek. I remembered walking to school, and the school friends I had left behind. I remembered learning to ride my bike with my brother holding the seat to give me confidence. The sudden thought of my big brother brought a tear to my eye, and a deep sadness suddenly settled over me and crept into my bones. I wandered along to the village church and searched through the cemetery until I found his grave. My only brother had died in his teens when he fell through the ice on the loch. Through my tears, I was shocked to read my mum and dad's name's on his tombstone. I fell to my knees on the cold wet grass and my tears became a flood. I had left it too late. When I could finally read the headstone, I saw that they had been dead for seven years. They had died within a month of each other. I sat by their grave and wept for the long ago past, for my brother, for my mum and dad, and when I rose I felt a vast emptiness. For now, I was truly alone. Everyone I had ever loved was dead. I now desperately wished that I had visited them, or even wrote, but it was too late now, and like my Sandy, they were gone forever.

I slowly wandered into the church hoping to find some comfort, but when I entered, I found it cold damp and empty. I looked around and realised the church was much smaller than I remembered, and I felt a strange coldness creeping into me in the echoic silence of the building. The wooden pulpit stood before me and I thought about the boring old minister whose sermons had droned on forever. My eyes fell on a memorial plaque on the wall and I read his name and immediately felt guilty for my disrespectful thoughts.

I wandered back to the Inn in a daze realising that there was nothing for me here. No one from my past, only strangers, only memories. I sat in my garret room and counted out my money. The thought of going home to an empty flat in Glasgow filled me with dread, so I decided I would stay for a few more nights.

That night the Innkeeper asked if I was going to the ceilidh in the village hall.

'No' I replied, 'I don't think so.'

'Go on hen it'll do you good,' the barman insisted, 'here you can have my ticket.'

'Aye come on,' his wife said, 'I'm going, so come and keep me company.'

Well I went to the ceilidh and I confess I did enjoy the entertainment. But I kept wishing my Sandy was sitting by my side, and holding my hand. After the Ceilidh we returned to the pub and over the next half hour, most of the people from the hall entered by the back door, and were now sitting around drinking and talking.

An old man came over and sat by my side. I had seen him talking to the innkeeper and saw them both looking at me.

'Did you say your maiden name was Young?' the old man asked.

'Yes,' I shakily replied.

'And was your fathers name Alexander Young from the village here?'

'Yes.'

'Man I knew your father well, and many a dram we had together, he was a fine shepherd, aye and a fine man. We both worked for the Laird at one time and we both did our time in the army together. I was best man at his wedding you know.

The old man told me that he had only recently returned to the village and that the place was now full of strangers.

'Now lassie, what brings you back here, and as the evening wore on I told him about Sandy and our life in the city.

'Och,' he said when my story wound to an end, 'I can understand why you wanted to leave the city; it's no place for a country lass, no place atall.'

At midnight, the pub suddenly became quiet and the barman turned off the lights. Shortly after we heard a loud cough from outside, and the door handle shook. Then we heard loud footsteps walking away. Five minutes later, the lights came back on, and the old man explained that it was just the village bobby doing his duty.

'He will be back just as soon as he gets out of his uniform,' he informed me. And sure enough, a wee while later I saw a tall stout man with shiny boots enter from the kitchen, pick up his half-full beer glass from the bar and resume his conversation with the barman.

Someone began to sing 'Amazing Grace,' and as the voices in the room faded. I looked around and saw the singer; she was the wife of the bartender.

Her singing was soft and sweet, and as she sung she looked at me, she was singing the song that I told her was sung at my and Sandy's wedding all those years ago. Her voice brought back a flood of memories and I began to cry.

When the song ended, everyone applauded and as she sat down someone else stood up. He sang a lively song about fishermen and the sea. As the dark night wore on and the singing continued, I felt some of my sadness leave me.

The barman was the last to sing and his gentle tenor voice sang a song I recognised; it was one of Sandy's favourites, 'Kishmulls Galley.' When the applause died down, he looked at me and asked if I would like to contribute to the entertainment, and egged on by the old man and the company, I nervously got to my feet and recited the only thing that I knew well, one of Robert Burn's best-known poems, 'Auld Lang Syne.'

When I sat down, the background noise slowly returned. A man came over to our table with three plates of stovies.

'Hi Dad,' he said to the old man, and to me he said, 'it's a while since you and I walked to the school together Margaret,' I looked at the tall handsome man as he handed me a fork and saw only another stranger.

'Don't say you don't remember me.'

He held out his hand, 'John Rose' he said and I remembered the name and remembered the boy that went with it, spotty Rose, a picture of a thin gangly boy with his head stuck in a book came into my mind. As we ate our stovies, we talked about our school days. His memory was much better than mine. He could recite the names of all our old classmates. An hour later, I excused myself and climbed the narrow stair to my lonely room.

As I lay in my cold bed the sound of people singing Christmas carols drifted up from below and added to my feeling of loneliness and sadness, and I once again cried myself to sleep.

Next morning when I came down for my breakfast, I found John Rose sitting in the small dining room. When he saw me, he stood up and wished me a merry Christmas.

He saw the look on my face and said, 'don't tell me you forgot it was Christmas as well?' I had forgotten it was Christmas, 'did you not hear the carols last night?' Suddenly I remembered the bottle of malt whisky I had hidden in the cupboard. This was Sandy's Christmas present from me; I had saved up all year for it, a few pence at a time. Suddenly I felt the tears in my eyes again.

John took my hand and led me to my table, he handed me a napkin and I dried my eyes.

'Sorry' I said, trying to smile, 'I forgot.' My breakfast arrived and John had a cup of tea, he told me he didn't stay in the village now, and that he had an office in the town.

'Your office,' I enquired, 'what do you do?'

'I'm a solicitor,' he informed me, 'and I came here last night to see you before you disappeared again.'

'How did you know I was here?' I asked in a puzzled voice.

'My dad phoned me last night to tell me you had turned up, and I drove down right away. I was coming here this morning anyway to spend Christmas with him.' I sat staring at him wondering why he would want to see me.

'It's about your parents' he said, 'I have been trying to find you for quite some time but no one knew your married name.'

'Well you found me now; did my parents owe you money?' I asked remembering the granite headstone on their grave. John laughed, and reached out and took my hand.

'Oh no,' he said, 'I was your parent's solicitor. Now Margaret I have some good news for you. First, your parents left you their cottage, so it belongs to you. I rented it out to stop it deteriorating, now if you want to move back to the village I can serve notice to the tenants and you should get possession in two or three months.' I sat in silence listening to this unbelievable news.

'There's a little more to tell yet,' John said, 'did you know your father had a brother in Canada?'

'Yes, Uncle Mike, he visited us once when I was a wee girl. I remember he taught us how to make paper aeroplanes.'

'Well,' John replied,' he died about ten years ago and left your father 70,000 Pounds in his will.'

The events I have just told you about happened over two years ago. I moved back to the village and John and I became good friends.

Then just three months ago John held my hand in both of his, looked into my eyes and he told me that as a boy he had a terrible crush on me. Perhaps that's why I never married, he added with a smile. Then he got down on one knee and asked me if I would become his wife. I found it a hard decision, and asked for a few days to think about it, for I considered myself already married to Sandy. The next day I returned to Glasgow and sat down by Sandy's grave side, I told him about John, and asked if he minded if I got married again, but the cold grave gave me no comfort, and I realized the decision was mine and mine alone. I loved Sandy with all my heart and he loved me. Then suddenly I heard Sandy's words in my head. If anything happens to me Margaret, don't you go pining your life away now. I remembered he said the words years ago, just before the army sent him to Ireland.

I returned to the village feeling that I had Sandy's blessing and accepted John's Proposal.

End

John and Margaret enjoyed a plate of stovies in the pub. I've mentioned them twice now so you really must try making them. Stovies used to be served in the pub, at the village dance or around the curling pond on a cold winters day. They are quick and very easy to make and a complete meal.

Stovies
One handful of diced precooked mutton per tattie. (Potato)
Three of four medium sized tatties, peeled and sliced to varying thickness.
One onion, thinly sliced, per two tatties.
A tablespoon of dripping, or best of all the jelly from the cooked mutton, if not available just use a little cooking oil.
Stock made from the mutton bone, or water works just fine.
Salt and a little pepper to taste.

Melt the dripping in a large pan, or use oil, add the ingredients, a layer of potatoes, onion, meat and so on. Cover with stock, or water. Not too much or the stoves will be soggy. Cook over a low to moderate heat for about half an hour. Shake the pan occasionally to prevent the bottom layer from burning. When the potatoes are soft to mushy, the stovies are ready. Invite me for supper and serve with a few homemade oatcakes.

Kishmul's Galley, sung in the pub is a (waulking) song from the Isle of Barra. Translated from the Gaelic. Waulking is a process of 'fulling' Harris Tweed, to make it wind proof.

Kishmul's Galley.
(Anon.)

High from the Ben a Hayich.
On a day of days.
Seaward I gazed. Watching
Kishmul's galley sailing.

Homeward she bravely battles.
'Gainst the hurtling waves.
Nor hoop nor yards.
Anchor, cable nor tackle has she.

Now at last, 'gainst wind and tide.
They've brought her to.
Neath Kishmul's walls,
Kishmul Castle, our ancient glory.

Here's red wine and
feast for heroes.
And harping too.
Sweet harping too.

The castle of Kishmul stands on a small rock island in Castle Bay Barra. Legend tells us, that a thousand years ago, this castle was the home of a pirate raider. Kishmul possessed a war galley of the Norse design and carried a crew of sixty-four fighting men. They raided, and plundered the Hebrides and across the Minch to Mainland Scotland and south to Ireland.

Trifle this is a pudding without equal. There is no doubt in my mind that Scottish trifle is the very best 'pudding' in the world. I have had pudding called 'trifle' in many countries, but they all fade into nothingness compared with the real thing.

Scottish Trifle.
Bake a sponge using queen cake recipe
Make ½ pint of yellow custard or buy both.
Spread sponge with raspberry jam and cut into cubes.
Place in a large glass bowl.

Use some of the juice from a tin of peaches (or other fruit) and mix with a large glass of whisky, 'bottoming' and sprinkle over the sponge until moist.

If you're a wimp use sherry.

Make jelly, using wine and rest of the fruit juice. Use 'Chivers' jelly cubes or similar, then pour over the peaches.

When set pour on the custard, when that is set, add whipped cream and sprinkle with hundreds and thousands or grated chocolate. If you want to waste this pudding by giving it to the kids, it might be an idea to leave out the whisky.

<u>The White Horse Treasure</u>.

Many years ago when I was a lad of fourteen, I accompanied my parents to the sleepy Moray Firth town of Nairn, for the summer holidays. As it was my birthday my parents indulged my passion for boats and fishing by booked a fishing trip with two local worthies, David and Eric Main.

The brothers at that time must have been in their seventies or perhaps even older. Both bearded men dressed alike in black pea jackets, black trousers and wore peaked skippers hats. The brothers owned an old wooden fishing boat, also black; with lines of tar between the deck planking, that oozed out and became sticky on hot summer days.

They made their beer money by taking visitors, like me out for a days fishing. We would leave harbour on the tide and steam across the firth to one of the many fishing marks known only to the men. That particular day, the day I remember well, we fished on a sunken wreck below the twin Souter hills, which guarded the entrance to the Cromarty firth.

It was on this trip, that Eric mentioned something to his brother that has influenced my life to this day. The old boat was silently drifting on a still, glassy sea. I stood with a fishing rod in my hands as I idly watched the bottlenose dolphins jumping, breaching and generally splashing about. The brothers stood in their wheelhouse and I saw the odd puff of pipe smoke wafting out from the open window. I stood fishing below the same window listening to the men talking and generally lugging in to their private conversation.

'David,' Eric had said, in a low conspiratorial voice that made my ears prick up. I quietly stepped onto an up-turned bucket the better to hear.

'Aye Eric, what is it?' his brother replied.

'Well' he said, 'I was just kinda wondering if you had any of yon white horse treasure handy.'

I was really concentrating now; he said the magic word 'treasure.'

'Why do you ask Eric? Are you in need?'

'Well since you ask I'll tell you,' he replied, it's...'

Just then, a big cod took my bait; caught off balance by the pull I fell off the bucket and crashed to the deck. The galvanised pail shot away, rolling and clattering across the deck. It's the first time, in fact the only time I ever regretted hooking a fish. Needless to say, the brothers ceased their private conversation and came out of the wheelhouse to see what all the commotion was about.

The white horse treasure, he had said the words as clear as day. Now I suppose I was a little on the cocky side at that time, and thought that all I had to do was ask, so ask I did.

'What's this about treasure?' I blurted out.

Both men turned and looked blank.

'Treasure,' I repeated, 'I heard you say treasure.'

'Och dinna be daft boy, away you go back to your fishing, and stop you're daydreaming.'

'But,' I persisted.

'What's wrong with you boy?' one of the brothers shouted, 'can you no take a tellin?'

'Away with you now and stop you're pestering,' the other added. Realising that my chance of learning more was now extremely unlikely. I went back to my fishing as ordered, but my heart was no longer in it. Not only had I lost the cod, but my mind was in turmoil. I thought about my half-read book, 'Treasure Island,' lying by my bed in our digs. Treasure, treasures, gold doubloons, long John Silver, and pieces of eight squawked by disreputable parrots. These are the thoughts that filled my head.

The town of Nairn, where the brothers kept their boat, was once described as having the longest street in the world. So long in fact that the populous at one end of the street could not understand the people who lived at the other. This story was based on fact, for at one time the fisher folk only spoke the Gaelic, whereas in the main town only English was spoken. Mind you, that was a long time ago. Now Jack and Eric were fisher folk, part of that now dwindling close-knit community that had always kept very much too them selves.

That long ago snatch of conversation took me the best part of thirty years to resolve. When I did hear the story, it was from a man who belonged to another fishing village along the coast to the east. His grandfather came from Nairn, and the story was passed down to him by his father.

This is what the man told me.

After the battle of Culloden in 1746, Bonny Prince Charlie fled from the Highlands, back to France. Meanwhile the Redcoat army took their bloody revenge on the Highland people. Whole families were put to the sword, crofts burnt and hundreds made homeless, and hundreds of good men were hung on the government gallows. To make matters worse famine gripped Scotland in its hungry grasp.

Many a family fled overseas during these troubled times, most never to return to their lonely villages and crofts.

Those who stayed behind had to conform; the wearing of tartan plaidie, or kilt was banned, the carrying of the dirk or sword forbidden, and an attempt made to destroy the clan system. But worst of all, the native language of the Highlands and islands was no longer acceptable, English was to be spoken, only English, and English history was to be taught in the school. The Gaelic was banished to the far corners of our once great nation and the soft words no longer spoken in public. New forts were built all through the land to enforce the new order.

Now one new fort on the shores of the firth was a vast fortified star shaped construction, using the very latest design. The site chosen was on a peninsula with the sea on three sides adding to its invulnerability. There was a weakness however, a small weakness, not so much to the fort itself but to its supply lines. The only problem was the lack of an army to exploit this weakness. Time passed and the vast fort slowly took shape.

A small group of men sat in a smoky fisherman's cottage, they spoke in their native tongue, the only language many of them could speak. They had a plan, a daring plan, failure would mean certain death but success could bring wealth and freedom. Freedom to start a new life here, or in a far off land over the seas. The plan was discussed on and off for several months, all were keen enough, but they lacked a leader, a man who could plan and a man who could inspire them.

This man, unknown to the schemers, lived among them. He was known as a quiet man, a man of few words, a loner perhaps. He had married a local lass, who was said to have refused to leave her family and friends. So he had moved and made his home in Fishertown. He worked part time on his wife's father's boat with her brothers. He bred dogs and made dog-buoys * from their skins, and he repaired nets and worked at the lobsters. It was a hard life but he made ends meet, most of the time. It was by accident that his name came to the schemers ears. All knew him of course but no one knew him well, or for that matter his history. He was not the sort of man you would ask.

It seems that this man's wife had got involved in a heated argument with her fellow workers. They had been hanging the herring on wooden rails in the smokehouse at the time. The subject of the disagreement is lost in time but some of what was said was still remembered. The wife it appears let slip that her husband had been a leader, a commander no less, and during the recent uprising it was he who routed General Johnnie Cope, and his Hanoverian army and sent them back south with their collective tails between their legs.

These were dangerous words; if the Redcoats heard them, he would be arrested and more than likely hung. It was a wall well known fact that thousands of men had already met their end at the end of a government rope for taking part in the recent rising. Anyway, the Redcoats never heard, so he was safe, well from outsiders anyway.

An approach was made and he was invited to a wee gathering, a ceilidh so to speak. The quiet man attended out of curiosity, and sat leaning forward with his head down and his elbows on his knees. He held the large dram he had been given in his big-calloused hands and listened in silence as the schemers told him of their plan. When the words stopped, the room grew quiet. He lifted his head and stared at the men. His menacing look made the schemers afraid and none wished to be the first to break the silence. John slowly rose to his feet and stood to his full height of six foot two. He downed his dram and wiped his bushy red unkempt beard with the back of his weather-beaten hand. His piercing wild eyes swept the small room, pausing and staring at each man in turn; they all without exception dropped their eyes to the floor. Then he flexed his broad shoulders and stepped forward. John was an intimidating sight, and not surprisingly, those present feared him. He stepped over to the table and noisily slammed down his now empty glass.

John then turned and walked towards the second biggest man present, the man who had been given the dubious honour of inviting John to this meeting. This man who had been leaning against the wall stood up to face the approaching giant who stopped almost standing on his toes and looked down at him. The quiet man asked in a low voice filled with menace.

'What makes you lot think I'd be interested in your daft scheme? And why did you ask me, to your meeting?'

'Now, now John,' one of the braver schemers who sat at the table said, 'we asked you here as a friend, all those present fought for the Prince, and many of our friends lie on that bloody moor, un-avenged. We know what you were,' the speaker continued quietly, 'and we asked you here as a friend and as a fellow Jacobite, to help us do this thing. We are all willing to take the risk, but we have always been followers, we need a man of experience to lead us.'

'What you ask,' John replied to the seated man, 'will bring danger to us all. Our enemies stay at the other end of King Street, perhaps even closer.' He said, looking around the small room, 'and as you know full well, the town is swarming with Redcoats.'

'Now John we are all friends here and loyal men…'

'Who told you about me? And who else knows?' John interrupted in the same low menacing voice.

'Now John, you might as well know that it's fairly common knowledge in Fishertown who you are, but rest easy man, your secret is safe here.'

'Thank you for the dram,' John growled, 'and now I'll bid you all a good night.' With that, he turned on his heels and ducked out the door of the smoky peat stained bothy.

'That man,' one of the schemers said breaking the silence of the now seemingly empty room. 'I was just wondering if he had any help with him when he routed Johnnie Cope's army.'

'Aye' said another; 'if we could only convince him to join us we could not fail.'

John walked home through the dark narrow lanes that twisted and turned between the tightly packed houses. He was angry, he had come to Fishertown because he was unknown here, now that had changed and he knew he would have to move again. He always had a hankering to go to Canada, or even South Africa. Perhaps, he thought, this would be as good a time as any to go.

Several days later, one of the schemers, the seated man who spoke to him at the meeting, met with the quiet man. Perhaps by chance, perhaps not.

'Well John,' the schemer said, 'it's nice to see you looking so well; and how's the fishing going?' John looked at him and the other took an involuntary step back. 'I'm just asking out of politeness you understand, no offence meant atall.' (at all)

'And none taken,' John replied and the other man visibly relaxed.

'Yon plan of yours you mentioned the other day.' John continued, 'I have given it some thought, now I'm not saying I'm interested mind you, but perhaps we might discuss it a wee bit more, just by way of conversation you understand.'

The wee discussion took place the following night and John took charge from the start, he stood and looked at the men present, he now knew the name of each one, and he knew just who they were.

'What you plan to do,' he said slowly, 'is to relieve the Hanoverian army of their wages. That is simple highway robbery. The money will all be in coin, and be very heavy,' he added. 'It will be well guarded by highly trained soldiers and if you succeed in relieving them of this money, all Hell will break loose.' The schemers sat in silence, unsure whether or not John was with them.

'Now let me list what would need to be done. First fighting men, perhaps as many as thirty, and you would have to arm them. Next, you would have to prepare a safe escape for those who want to leave, that is, before the theft is discovered. Those who want to stay will need a very good alibi.'

One of the schemers stood up, 'with your permission John I can answer some of your questions right now.'

'Go ahead David Main,' John said.

'The weapons are readily available, I know of the whereabouts of fourteen Brown Bess muskets, twelve pistols and enough long swords to arm another rising.'

'What about powder?' John asked.

'Well there you have me,' David replied scratching his head,' I have lots of flasks but not much in the way of powder to fill them. We might manage to charge the muskets once but that's all. We do,' he said in a rather cheerier voice, 'have a few stone of lead put by for bullets.'

'Thanks to the Kirk roof no doubt,' John replied with a grin.' I heard it was leaking. Most of you here are fishermen,' John continued 'and therefore have access to boats, so that may solve some of your problems.'

During the next hour, the men told John what they knew and what they hoped he would help them do. John remained quiet and studied the men present, when the last man stopped talking all the eyes returned to the silent man.

'You have observed well,' he told them, 'you know that the soldier's wages arrives from the south every six months and you know what road they take and you have counted the escorting redcoats. The next shipment you say is due in six weeks time. If we,' he used the word 'we' for the first time and those present opened their eyes a little wider and a faint smile played on their weather beaten faces.

'You're with us then John?' someone interrupted.

'Aye it looks like it, but if we are going to do this thing then its best done soon, the longer we wait the more danger there will be. As far as the men and gunpowder are concerned, I can help; I can also supply a lot more muskets. After we do this thing I will want to leave right after with my share, and I'll no be coming back. I'd advise you lot to do likewise.'

The meeting wore on and by midnight it was all agreed, they would do it.

Over the next weeks the planning continued, muskets were recovered from the thatch roofs, swords dug up cleaned, polished and sharpened. Then one day they visited the campsite the soldiers always used and John examined the topography in great detail. He sat on his own on top of the only high ground, the top of a sand dune that overlooked the site. Several times during the next hour he got up and wandered away to look at some feature or other, then he returned to his high ground.

When he called his group together, he looked relaxed, for now he knew how it could be done. They set out towards the coast and found a way over the sands of Culbin. The desert stretched for over twenty miles along the coast and six miles inland. They skirted around the small isolated villages and crofts that lay scattered along the edge of the sands and eventually sat down overlooking the sea.

'It would just be possible,' one of the fisherman said, 'to bring a few fishing boats close in to the shore behind that sand bar out there, that is if the weather was kind and we had a high tide.'

'Try it,' John ordered, and find out what can be done.'

John then went away on his own for two weeks. 'He's gone to visit a dying relative who might have a boat to leave,' was the story told by his wife. It was a good story, and more than satisfied the curious.

During his absence, the schemers carried out Johns instructions. They took one of the boats along the coast and sounded the depth. Eventually a deep channel was found that led in behind the bar. They also discovered a 'hole,' a few boats could lie afloat in at low water; and escape again at half tide. They then cut and delivered several cartloads of dry firewood and left it in the exact spot John had marked.

The day at last arrived and all were ready and eager. With John as their leader, they believed they could not possibly fail. Three fishing boats lay at anchor behind a sand bar ten miles to the east of the town, on board were twelve women and sixteen children. They had boarded the boats under cover of darkness carrying their few pitiful possessions wrapped in cloth. The wind was out of the west and the sky told of bad weather to come. Ashore twenty-six men gathered. Some were men who once served under John. The rest came from Fishertown or men from the surrounding area.

Some of the men had marched over night to the chosen site, while others had travelled for many days.

John deployed the men with great care, taking each in turn to his post, ensured that they were completely hidden and that the two muskets they carried were loaded with dry powder. Finally, he confirmed that they each knew exactly what was required of him. The men had a score to settle with the hated Redcoats. And tonight they would take their bloody revenge.

Three hours later, as darkness began to descend, the sound of horses and the jingle of harness was heard. John's men settled into their cover and sat tight.

As the redcoats came into sight an officer barked an order, and the soldiers went about setting up the camp. Instead of foraging for firewood, they stole wood from the stacks of dry sticks Jocks men had left handy for them. They were also delighted to discover several jugs of whisky, hidden in the woodpiles.

Soon the smell of cooking filled the air. The Red coats fed the already big fires and greedily consumed the good whisky. Men were talking loudly, singing and Jock heard an officer laugh, at what was likely to be his last joke. And as the night wore on the camp fell quiet, the majority of the soldiers stumbled to their beds, only the unlucky ones chosen for sentry duty remained; they were barely awake.

Now and again, a grunt or a groan was heard out in the darkness but it was ignored. The hoot of an owl coming a second later distracting the listener from the sound he should have heeded. John's men had taken out all the sentries, without a shot being fired. Now they crept slowly towards the officer's tent.

Suddenly a shout rang out, followed by a musket shot. As the camp came alive, lead filled the air. The soldiers were caught in the bright lights cast by the still blazing fires. The deadly crossfire that came out of the darkness ripped the confused men to pieces. They had no hope. John had planned it that way. Within a minute, the last shot had been fired.

In the silence that followed, the odd pitiful sound of a wounded men could be heard, or their scream as they saw the swing of a long sword above them. Soon it was still and the silence became profound.

The murderous nights work was over; now all they had to do was escape. They quickly hitched the horses to the wagons, and they set off leaving the dead for the crows, who would no doubt find them by breakfast time.

The men streamed out ahead, marking the trail. Then they entered the sands and wound their way through the high dunes. The wind was stronger now and the airborne sand stung their faces. They were a mile from the coast when the wagon sank down to its axels in the soft sand. Forget it. John ordered. Each man take what he can carry, and go. The wind continued to build and the fisherman feared for their boats and families. They headed for the beach and threw what they carried into the waiting rowboats and headed out into the approaching storm. Those who wished to remain stumbled home along the shore laden down with their ill-gotten gains. The quiet man slipped away into the night with his wife and no one knows if he reached Canada or not. Some say he did, and that the quiet man became a teacher. As for those who chose to stay few survived, for as John predicted, all Hell broke loose. As to the bulk of the money, that was lost under the shifting sands.

A hundred years passed by. A mere tick in eternal time, but the sands had moved on. Another small croft now lay in the path of the marching dunes, and in the nearby village, most of the people had already left. Only one sad lone crofter remained. He was reluctant to leave the home he was born in, the home of his father and his only means of support.

A white horse stood by, as the defeated crofter looked on. His master's home was half-full of sand and dwarfed by a huge sand dune, his stable was gone, and within days, the field that grew his hay would also be gone.

The crofter had decided to visit his sister, and perhaps ask for a job on one of his cousin's boats. The white horse shook its head and the sound of jingling harness drifted over the sand like an echo from the past. The sky grew dark, and a sudden gust of wind stirred the crofter into action. He glanced at his few possessions on his cart and wiped a tear from his eye. The airborne sand began to sting his face, so he moved his jacket and placed it over his horses head to protect the animal's eyes. He then took hold of the bridle, and moved out into the sands, heading for Fishertown.

The wind increased and within an hour he realised he was lost. The defeated man trudged on picking his way through the moving sand dunes. He navigated by instinct, heading he hoped inland. The white horse plodding by his side, then as they wound their way around a tall dune the white horse stumbled. As if on cue, the wind suddenly died away and it grew deathly quiet. The crofter lifted his head and his gritty eyes fell on the remains of a four-wheeled cart that had just emerged from the sand. By his horses feet lay a steel bound chest surrounded by the glint of silver coins.

End

Although my story is fiction, I know for a fact that the White Horse treasure, did, or does exist. I have also been told stories of ships cargoes that were lost in the sands.

Culbin Sands. John Martin of Elgin describing Culbin during a 17th century sandstorm.

The wind comes rushing down through the openings between the hills, carrying with it immense torrents of sand, with a force and violence almost overpowering. Clouds of dust are raised from the tops of the mounds and are whirled about in the wildest confusion and fall with the force of hail. Nothing can be seen but sand above, sand below and sand everywhere. You dare not open your eyes but must grope your way about as if blindfolded.

Excerpts from "A Short History of the Scots (written by Harry Kinnaird - 02)

Culbin

'The Barony of Culbin was lost to the Kinnairds after the disaster known as the mystery of the Culbin Sands, when the estate became overblown by sand from the Moray Firth and the rich farmlands were destroyed. This was reputed to have happened overnight, however this is improbable and the destruction was more likely to have been progressive culminating in 1694/95 when the Laird's house and remaining land became buried by sand and unproductive. This was a sad end to an estate which had had 3600

acres containing sixteen farms and was considered rich enough for Montrose to plunder in 1614, fifty years earlier.'

Dog buoys.
You won't like this, but I'll tell you anyway. Once upon a time dogs were bred for their skins. Their skins have no pores, so were therefore ideal for fishing buoys.

A wee bit of history.
This is a good place to insert a wee bit of Scotland's history. No, it's not boring but I will keep it short. I'll also give you the words and the background of a few Scottish songs.

The past, as recorded in textbooks and taught in schools, seldom gives the view of the ordinary people. In many cases, it can be blatantly inaccurate and misleading. The history recorded in songs and verse, on the other hand, was normally written by, 'the ordinary people,' who it could be argued, were equally biased.

For those of you who play an instrument, I have included the guitar cords I use for some of the songs.

This small collection of songs gives a brief and by no means full account of events covering a period of about fifty years. Starting with the Treaty of Union 1707, and ending shortly after the Battle of Culloden 1746.

History in songs.
The first song I have chosen was collected and re-written by Robert Burns. For those who don't know who he was, he is our Scottish Bard. (Poet) (1759-1796)

The song tells the story of the 'Treaty of Union'. When, 'a parcel of rogues,' in the Scottish Parliament, voted into law the effective destruction of the Scottish nation.

Scotland and England then united under one Parliament, to establish the United Kingdom of Great Britain.

Such a Parcel of Rogues in a Nation.
By Robert Burns

(Em) Fare weel tae a' oor (G) Scottish fame,
fare weel (Am) oor (Bm) ancient (Em) glory.
Fare weel e'en tae the (G) Scottish name,
sae famed (Am) in (Bm) martial (Em) story.

Noo (G) Sark (D) rins ower (Em) the (G) Solway sands,(Em)
an' Tweed (D) rins tae the (E) ocean
Tae mark (G) whaur (D) England's (Em) province (C) stands,
such (G) a parcel o'(D) rogues in a (E) nation.

What force or guile could no I subdue, thro many war like ages.
Is wrought now by A coward few, for hiring traitors wages.
The English steel we could distain, secure in valour's station.
But English gold has been our bane, such a parcel o' rogues in a
nation.

O would or I had seen the day, that treason thus could sell us.
My auld grey head had lain in clay, we Bruce and loyal Wallace.
But pith and power till my last hour, I'll mak this declaration.
We're bought and sold for English gold, such a parcel o' rogues in
the nation.

(Charlie) Charles Edward Stuart, or Bonny Prince Charlie, was the grandson of the deposed, King James 11, (1633-1701) and son of James, the Pretender, who failed in two attempts to regain the Scottish and English thrones. This song tells of the encouragement given to the Prince to come over to Scotland.

Come Ower the Stream.
By, James Hogg. (1770-1835).

Come ower the stream Charlie,
Dear Charlie, brave Charlie,
Come ower the stream Charlie,
And dine wi' MacLean;
And though you be weary
We'll mak' your heart cheery,
And welcome our Charlie;
And his loyal train.

We'll bring down the red dear
We'll bring down the black steer,
The lamb from the peared
And doe from the glen.
The salt sea we'll harry,
And bring to oor Charlie,
The cream from the bothy
And curd from the pen

And you shall drink freely
The dews of Glen Sheerly
That stream in the starlight
When kings dinna ken;
And deep be your meed.
o' the wine that is red,
To drink to your sire
And his frien' the MacLean.

If aught will invite you,
Or more will delight you
'Tis ready a troop
Of oor bold Hieland men
Shall range on the heather
Wi' bonnet and feather,
Strong arms and broad claymores,
Three hundred and ten.

This song spreads the news that the 'King' has landed on Scottish soil. Collected by Lady Caroline Nairn (1766–1845)

Wha'll Be King But Charlie?

The (Am)news frae Moidart cam (G)yestreen
Will (Am)soon gar manie (C)farlie
For (Am)ships o' war hae (C)just come (G)in
And (Am)landed (G)Royal (Am)Charlie

Come (Am) thro' the heather, around him gather.
You're (Am) a' the welcomer (G) early.
(Am) Around him cling, wi' (C) a' your (G) kin.
For (Am) wha'll be (G) King but (Am) Charlie.

Come (Am) thro the heather, (C) around him gather.
Come (Am) Ronald come Donald, come a the (G) peare.
(Am) And crown your peared (Am) peare (G) King!
For (Am) wha'll be (G) King but (Am) Charlie?

The Hieland clans, wi sword in hand.
Frae John o' Groats to Airlie.
Hae to a man declared to stand.
Or fa wi Royal Charlie.

The lowlands a, baith great an sma'.
Wi mony a Lord and Laird, hae.
Declar'd for Scotia's king an law.
An pear ye wha but Charlie.

There's ne'er a lass in a the lan.
But vows baith late an' early.
She'll ne'er to man gie her heart nor han.
Wha wadna fecht for Charlie.

Then there's a health to Charlie's cause.
And be't complete an early.
His very name our heart's blood warms.
To arms for Royal Charlie!

A farmer who witnessed the battle of Preston Pans, (1745) tells the story of the famous victory for the Jacobite army; when they fought the greatly superior Government forces led by Sir John Cope. The battle took place just outside Edinburgh.

Johnnie Cope By
Adam Skirving, (1719- 1803)

Hey, (G) Johnnie Cope, are ye wauken yet?
Or (D) are your drums a-beating yet?
If ye (Em) were wauking I (G) wad (D) wait.
To (Em) gang to the (D) coals in the (Em) morn-in.

Cope (Em) sent a letter frae Dun---bar,
Saying' (D) Char---lie, meet--- me-- an' ye--- daur.
An- (Em) I'll learn (D) you the (Em) art o' (D) war.
If you'll (Em) meet me (D) I-- the (Em)morn--in.

When Charlie look'd the letter upon,
He drew his sword the scabbard from;
'Come, follow me, my merry men,
An' we'll meet Johnnie Cope in the morn-in.'

'Now Johnnie, be as guid' your word,
Come let us try baith fire an' sword,
An' dinna flee like a frichted bird,
That's chas'd frae its nest in the mornin.

When Johnnie Cope he heard o' this,
He thocht it wadna be amiss
To hae a horse in readiness,
To flee awa in the mornin.

Fye, Johnnie, noo get up an' rin,
The Highland bagpipes mak' a din,
It's best to sleep in a hale skin.
For 'twill be a bloody morning.

When Johnnie Cope tae Dunbar came,
They spier'd at him, 'Where's a' your men?'
'The deil confound me gin I ken,
For I left th em a' in the mornin.

Now Johnnie, troth, ye werena blate.
Tae come wi' news o' your ain defeat.
And leave your men in sic a strait.
Sae early in the mornin.

In faith,' quo' Johnnie, 'I got sic flegs.
Wi' their claymores an' phillabegs.
Gin I face them again, deil break my legs!
Sae I wish you a' gude mornin.

The Jacobite army marched south, fully expecting the English and French Jacobites to join them in their cause as promised. But when the support failed to materialise, the Prince, and his army returned to Scotland.

On the 16th of April 1746, the Jacobite army suffered their only defeat when they met the Hanoverian army; whose ranks were swelled by Scottish clans opposed to putting a Stuart on the throne.

The battle of Culloden was the last battle fought on Scottish soil. The Jacobite army numbered about 5,000 men, and the Hanoverian army, supported by about sixty percent Scots, numbered about 10,000. The battle took place above Inverness on Drumossie moor or Culloden. Some Clans, who were unsure which side to fight for, stood by and watched.

The aftermath of Culloden brought bloody and indiscriminate reprisals throughout the Highlands. Over a thousand men met their death on the government gallows.

The Disarming act, passed by the London parliament, made it an offence to carry even a dagger in self-defence and the wearing of the plaidie or kilt became a crime punishable by deportation. Throughout the Highlands, the homes of the guilty and innocent burned and people turned out of their crofts and farms. Thousands emigrated, leaving Scotland never to return.

This is one of the few web sites that have gone to the bother of looking up the facts, and is well worth a visit. <www.historichighlanders.com/whyculloden.html.> Click, Why Culloden.

Remember when reading, history and not only from this period, that many historians use the word English army, or English, when they mean British. It is sad that many people think that Scotland Wales and Northern Ireland are English counties.

According to this song, Bonne Prince Charlie escaped to Skye dressed as Flora MacDonald's maid Betty. The last verse tells of the beginning of the bloody reprisal's the government ordered agents the Scottish People.

Skye Boat Song.
By Sir Harold Boulton, Bart.
(1859-1935)

Chorus
(C) Speed, bonnie boat, (Am) like a bird on the (G) wing,
(C) Onward! the (F) sailors (C) cry;
Carry the bairn (Am) that's born to be (G) King
(C) Over the (F) sea to (C) Skye.

Loud the (Am) winds howl, loud the (Dm) waves roar,
Thunderclaps (Am) rend the air;
Baffled, our foes stand by the (Dm) shore,
(Am) Follow they will not (G7) dare.

Though the waves leap, soft shall ye sleep,
Ocean's a royal bed.
Rocked in the deep, Flora will keep
Watch by your weary head.

Many's the bairn fought on that day,
Well the claymore could wield,
When the night came, silently lay
Dead in Culloden's field.

Burned are their homes, exile and death
Scatter the loyal men;
Yet, o'er the sword cool in the sheath
Charlie will come again.

It is widely believed that this very popular song was written by a Jacobite prisoner, the day before he faced the government's gallows. The low road is the path of ghosts and spirits.

Loch Lomond (anon)

Chorus
(G) By yon bonnie banks, (E) and (C) by yon (D) bonnie braes
Where (G) the sun shines (Em) bright on (C) Loch (D) Lo-mond
There (G) me and my (Em) true love (C) spent many (D) happy days
On the (G) bonnie, bonnie (C) banks o' (D) Loch (G) Lo-mond.

Oh you tak' the high road and I'll tak the low road
An' I'll be in Scotland afore ye,
But me and my true love will never meet again
On the bonnie, bonnie banks o' Loch Lomon'

Twas there that we parted in yon shady glen.
On the steep, steep side of Ben Lomon',
Where in purple hue, the hielan' hills we view,
An' the moon comin' out in the gloamin'.

The wee birdies sing, and the wild flowers spring,
While in sunshine the waters are sleepin'
But the broken heart it kens nae second spring again,
Tho' the waefu' may cease free their greetin'.

The next song, gives the Government or Hanoverian point of view. The original version was simply an attack on the Jacobites. Robert Burns re-wrote the song in 1791. I think it could be the first anti-war, or protest song.

Ye Jacobites by Name.

Chorus
Ye (Em) Jacobites by name, (C) lend an ear, (Bm7) lend an ear,
Ye (Em) Jacobites by (Bm) name, lend an (Em) ear,
Ye (G) Jacobites by name, (D) your faults I will proclaim,
Your(Em) doctrines I might (Bm) blame –(Em) you shall hear, you shall (B7) hear!
Your (Em) doctrines I (Bm) might blame – you shall (Em) hear!

What is Right, and what is wrong, by the law, by the law?
What is Right, and what is wrong, by the law?
What is Right, and what is Wrong? A short sword and a long,
A weak arm and a strong, for to draw, for to draw!
A weak arm and a strong, for to draw!

What makes heroic strife, famed afar, famed afar?
What makes heroic strife famed afar?
What makes heroic strife? To whet th' assassin's knife,
Or hunt a Parent's life, wi bluidy war, bluidy war!
Or hunt a Parent's life, wi bluidy war!

Then let your schemes alone, in the State, in the State!
Then let your schemes alone, in the State!
Then let your schemes alone, adore the rising sun,
and leave a man undone, to his fate, to his fate!
And leave a man undone, to his fate!

The next two songs by the author are about the Highland Clearances.

The Far Highlanders
The 'P' in Ptarmigan, is silent

(C) Dark (Am) Culloden---, (C) dark (Am) Culloden---, you (Gm)
share in our (F) pain---.
As do--- the (C) far (G) Highlanders'--- who (G7) bear--- a clan
(C) name. ---
(G) Around --- the (G7) scattered (Am) croft (C) stones, --- the (G)
ptarmigan (C) stirs. ---
(G) To the (G7) echo--- of the (C) pibroch, --- that (G) weeps ---
(G7) the sad (C) dirge. ---

The dark cloud that rose spread over the land.
Brought on by the cowards, traitorous hand
Broken promises and betrayal, led them this far.
As reward for loyalty, were scattered afar.

The seeds were sown, those treacherous years
Grew strong and flourished, in blood and tears
For the lands were cleared and her people deported.
Without trial or compassion, cruelly transported

The laird slept soundly, far away from the frost
As smoke billowed high over our ancestors croft
The invaders they came, in vast flocks of pure white.
While homeless farmers shivered and died in the night.

The wind howls like a banshee, o'er the moor.
As our ancestor once wailed on that far distant torr
Now the un-avenged ghosts hunt the bloody hand.
That cost them their faith, their lives and their land.

Two hundred thousand were driven from the glens.
From the braes, the sheilings, and the white capped bens.
Some were bound and shipped, to America shore
Some went to Canada and saw Scotland no more

In Australia, New Zealand, on that far southern shores,
And in Canada and America' the pipes play as before.
Our clansmen have climbed to the lofty heights
And Scots the world over stir to the sound of the pipes.

Dark Culloden, dark Culloden you share in our pain.
As do the far Highlanders who bear a clan name.
Around the scattered croft stones, the stirs. Ptarmigan
To the echo of the pibroch, that weep the sad dirge.

In memory of the people who were deported during, 'The Highland Clearances. Set to the ancient tune. The Eriskey Love Lilt.

A Bitter Man.
Far beyond, the blue horizon,
Away across, the stormy sea.
There's a land, that I remember,
The place that I'm, exiled frae. (from)

From our Country, we were driven,
Burning homes, we left behind.
Not a penny, in our pockets
Or any hope, of better times.

Cursed be, the landed gentry.
May their lives, be filled with pain.
I wish them nothing, but misfortune
And the end of, their likes reign.

Now in a harsh land I toil
Beggars wages, sorely won.
Tending strange plants* in foreign soil
From early morn till the day is done

Nothing in life ever changes,
Everything is still the same
Wealthy men, have all the power
But here they go by a different name. *

Cursed be the landed gentry.
May their lives, be filled with pain.
I wish them nothing but misfortune
And the end of their likes reign

Hunger is my constant companion
It breaks my heart that I must see
My family around the spartan table

Bread and dripping for their tea. (Dinner)

Now my life is near its end
Bitter irony follows me to my grave
For I once fought for king and country
Or was I just the landowners slave?

* In the Highlands landowners were, and are called, The Laird.

* Some Highlanders were deported to the USA to work in tobacco fields.

Many men who served in the British army, fighting for king and country, returned home to find their homes burnt and their families deported.

OK that's the last one. Most of the above songs are sung by Scottish folk groups, many are available on CD or you might find them on u-tube, but not the last two, they're new, but if you like it feel free to sing or record it on U-Tube. Don't forget to send me an e so I can give you marks out of ten.

Did you notice that only one of the songs was written at the time?

The Highland Clearances. Describes the most heinous crime from Scotland's past.
Of all the crimes inflicted on the Scottish Highlanders, the Clearances are without doubt the most heinous. Even to this day, its memory invokes great bitterness.

Between the years 1783 and 1881. The landowner's greed, and inhumanity to their fellow man, resulted in a documented eviction, of 170,571, people from their homes and lands but the true number is known to exceed this figure.

The seeds of this crime were sown in 1707, with the Treaty of Union, and as Burns, song clearly states, the treaty came about without the consent of the Scottish people. After the battle of Culloden, the wearing of the kilt was banned. Religion, as usual, played its part by stirring up strife between Catholic and Protestant, adding fuel to the conflict. Then the British government inflicted the, 'Heritable Jurisdictions Act,' on the Highland Chiefs, this act stated that, 'those who did not accede to British jurisdiction were to have their lands forfeited to the Government.'

Highland landlords complied with this legislation.

Landowners, 'some say as many as sixty percent,' deserted their people and moved to Edinburgh or London.

In the 18th century, many landowners reached a 'financial impasse,' when the rents from their tenants could no longer support their extravagant way of life. Around this time, demand for wool rose and its price tripled.

The landowners seized on this opportunity to regain their wealth, but in order to fully exploit the sheep they decided they had to clear the land of the people. They instructed their 'factors' to carry out this crime. At the height of the clearances, the landlords were evicting over 2,000 families, every day. Their houses were burned and the stones scattered. Some of the destroyed crofts had been in the same families for over 500 years.

Elizabeth Gordon, Countess of Sutherland (1765 – 1839) and her husband the Marquis of Stafford, (later honoured, and made 1st Duke of Sutherland) 'cleared' 15,000 people from their traditional lands to make way for 200,000 sheep. Many of their former tenants starved or froze to death among the ruins of their destroyed homes.

A statue of the Duke still stands in Sutherland as an insult to the Highland people he wronged.

I will tell you one more story about the clearances, and then return to Loch Ness Tales Legends and recipes.

This story was taken from the web site.

'1851. The clearance of Barra by Colonel Gordon of Cluny. The Colonel called all of his tenant farmers to a meeting to 'discuss rents,' he threatened them with a fine if they did not attend. In the meeting hall' over 1500 tenants were overpowered, bound and immediately loaded onto ships for America.

An eyewitness reported '...people were seized and dragged on board. Men who resisted were felled with truncheons and handcuffed; those who escaped, including some who swam ashore from the ship, were chased by the police...'

It must be said that the peoples of Ireland suffered a similar treatment under the landed gentry, in some cases worst but that's another story.

The Keeper.

The keeper returned from his search. 'Aye, you got him all right,' he informed the shooter.' I found the blood, and here's a tuft of his hair,' he added as he retrieved a few deer hairs from his pocket. 'I found the spot right enough but the cover is so thick I just could not quite put my hands on your buck. Not to worry I'll away home, collect the dog, and come back for it later. I'll drop you back to your car now so you can get back to the hotel for your bath and dinner. No, no it's no bother atall. I'll manage just fine.'

They wandered back to the Land Rover and after the Keeper returned the guest to his car, he went home for his tea.

'How did you get on,' his wife asked as he entered and handed over his tip.

'Not bad,' she said counting the notes, 'so he got one then.'

'Well not quite,' her husband replied, 'yon man couldn't hit a barn door at twenty feet, I'll away out tonight with the light and get him a nice one, just to keep him happy.'

'If he didn't get a buck,' his wife asked, 'how come he gave you a tip?'

'Och well you know how it is, it's best to keep them happy, he got one right enough; well he thinks he did and it's the same thing really,' he said as he sat down for his tea.

The next morning the gamekeeper collected the guest and took him to the larder, a fine big Seika buck was hanging from the meat hooks awaiting his inspection.

'Man you're a fine shot,' the keeper said, 'the deer only went a few yards. I must have almost stood on it yesterday; the dog found it right away, no bother atall. You'll no doubt be wanting the venison home with you; I'll just tell the wife to pop it on your note.' The guest was delighted with the buck, it was even better than he remembered. 'Well now it's the river you'll be wanting today, so we better get a move on. Your friend you say is meeting us at the hut. Yes, yes the river is fishing just fine, we had a man here last week took three salmon in two days all nice clean fish. The biggest was just over fourteen pounds.' They arrived at the river just as the other guest drove up.

The guest wasted twenty minutes of the best fishing time explaining how he shot a big buck, and then they took another twenty minutes setting up their rods and generally wasting time. Then they decided to have a cup of coffee before starting out.

Our keeper, if he had been on his own, would have fished the beat, landed any fish that were there and gone home for his breakfast by now. At last, they were ready.

'Well now,' the keeper said turning to the guests, 'one of you can start your fishing here and I'll take the other up and show him the next stretch. I'll call back in a couple of hours to see how you're getting on.' The guests decided among themselves that the new comer would accompany the keeper up to the next beat, so they climbed into the Land Rover and drove away leaving our guest to thrash the river to foam. The keeper then went home for his breakfast.

He returned a while later and stood watching in disbelief. The first guest stood more or less, where he left him, his rod lay on the bank and in his hands, was a bird's nest of tangled fishing line. Our keeper stepped forward.

'Well, well I see you have got a wee bit of a knot there, yes, yes it's easy done, you would have had a fish on no doubt, it sometimes happens but I suppose it got away. The guest somewhat embarrassed, nodded his head. Yes, yes just as I thought, you away and have your tea, and I'll sort this out for you. No, no it's no trouble atall.' The guest gladly wandered back to the hut for a cup of coffee. When he returned, the keeper handed him back his rod.

'I just put a new line on for you; the one you were using was damaged. No, no I'll just tell the wife to make a wee note and you can square up with me before you go home, no problem atall. I'll just away down stream now and see how your friend's getting on.' Off the keeper went stuffing the expensive double tapered salmon fly line into his pocket, a nice wee job for the winter nights in front of the fire, he thought.

The other guest was a much better angler and the keeper stood and watched him for a while.

'How is my friend getting on,' the angler enquired.

'Oh just fine, just fine, he lost a nice big fish just a wee while ago, so there are definitely fish about.'

'Oh good for him,' the angler replied, do you know what fly he was using.'

'Oh just the same one I gave him myself this morning.'

'I don't suppose you have another to spare do you?'

'Well now I might just,' the keeper said making a show of searching his pockets, and then he removed his hat, 'well just look at that, it was in my hat all along. No, no, let me do it I'll tie it on for you, man this fly is the business, I'm surprised it's not banned altogether.' The keeper somehow forgot to return the original fly to its owner. He would sell it, 'as the business,' to the next guest.

'Just you try that now. Look, look over there, I saw a fish move, he lied pointing to a bit of still water, man you have a really good chance now. I'll just leave you to it and get on with my work.' He wandered off to have a look at the river; then lay down on a grassy bank and soon he was fast asleep. When he returned to the hut, two hours later he found the anglers happily discussing the fish they had seen or lost.

'I'll call back in the afternoon just to see how you are getting on,' he told them.

Back home for lunch, his wife met him at the door, 'well,' she said using her usual comment, 'and how are our fishermen getting on.'

'Oh just fine,' he replied, 'it's just a pity there are no fish for them to catch, the water is far too low and it's a fine sunny day. Not to worry they seem happy enough. Oh by the way, he wants thon (that or yonder) venison. Oh and add a couple of salmon flies and a fly line to their note.'

When he went back mid afternoon, he found the fisherman sitting on the bank having their coffee.

'Well, well now, have you put the fish you caught out of the sun? Not a one, you don't say. Well, well I just don't understand it, but I suppose that's fishing for you. Never mind the day is young yet so you never know. Now then, about tomorrow, you have booked a wee bit of rough shooting, I was just wondering if you wanted to shoot a few clays first, just to get your eye in. Oh yes, most of the guests do, no bother at all, I'm glad to help; say about nine.'

Now if by chance you get no fish, I have two in the freezer you can take home with you, no, it's no problem at all, just you give me the same as the fish man gives me and I'll be more than happy.

Next morning they arrived eager, forgetting the no fish day of yesterday. Our keeper had set up the well-oiled trap behind a pile of straw bales. Our shooters knew not what awaited them. First, the set up was a little unusual. The shooters were perhaps a little too far back from the trap, and the trap had a slightly stronger spring than was normal. The clays were, shall we say, well used, and made by our keeper out of fibreglass, and therefore virtually unbreakable.

'Now then,' the keeper said 'I'll just fire a few clays for you, as soon as you can hit two out of three we will set out. No, no problem atall glad to help.'

After he had fired his fifty fibreglass clays, he opens a box of regular ones. Then he wandered over to the guests, who by now were rather sore and confused.

'What shells are you using?' he asks in a concerned voice. 'Oh man those are no good, no good atall. I've got a few boxes in the Land Rover. Ones I loaded myself,' he said and wandered away to collect them. 'Now you try these, they are a bit more expensive that the ones you are using, but man, they're so much better. Perhaps you might try standing just a wee bit closer to the trap. Yes that's fine.'

He returned behind his straw bales and adjusted the trap to its slower setting, then loaded a regular clay. When the first shooter shouted pull and fired, the clay went into dust within twenty feet, as did the next six. The day had now worn on and the morning was all but used up.

'Would you rather go out after you have had a wee bite of your lunch,' the keeper enquires, 'no, no problem atall. Say one o'clock, right you are, I'll see you then.' He left them to it and wandered home for his dinner.

'Well,' his wife asked,' how did the shooting go?'

'Och they wanted a wee shot at the clays first just to get their eye in. Put two boxes of clays on their wee note for me, and four boxes of my shells. Oh yes and I sold them a couple of salmon out of the freezer, so dig out two big ones for them'.

The afternoon shoot went well and the guests bagged six rabbits, two pheasants a partridge and three pigeons... As they returned to the road, the keeper stood by the open gate with his hat off waiting for his tips.

All in all the guests had had a good few days in the country. They were however rather surprised at the size of the 'wee note,' the keeper read out; but he was such a helpful man they just paid up.

Next morning our guest returned to work in the Inverness tax inspector's office, he sat down behind his desk and examined the keeper's wee note. Then he buzzed his secretary and asked her for our gamekeeper's tax returns.

End

Venison. Have you ever tried it?

We have three species of deer in the Highlands, Red Deer, Seka Deer, (Jap), and Roe Deer. The red deer is the biggest, and once inhabited the vast Caledonian forest that covered the Highlands. As an echo of that long ago past the land where deer stalking takes place, is still referred to as, Deer forest. The habitat of the red deer today is out on the open hills, bens, moors and high mountains.

For the table, I prefer hind, or female Red Deer meat. The stag can be a little on the, 'gamey' side for my taste.

The Seka, from Japan, is smaller than Red Deer. They prefer to live in open woodland or low-lying hills. The Seka and Red deer can, and do on occasion interbreed.

The Roe, is our smallest deer, it usually lives in thick woods and forests. All venison is low in fat and therefore a healthy food.

Recipe. Roast haunch of venison
Simple method, for a haunch, or tied up shoulder roast, of about 2 kg in weight.
1-tablespoon olive oil
A few slices of fatty bacon.

Pre heat oven to 160°c, Place roast on rack inside roasting pan. Brush with oil; Cover with bacon slices. Roast for just over an hour, or about 20 minutes per ½ kg, do not over cook. Test by sticking a knitting needle into the meat. Remove the roast from oven to rest for about 10 minutes. Slice and serve with gravy made from pan juices, serve with a spoonful of cranberry jelly, and of course neeps and tatties.

As an alternative, loosely wrap and seal in foil, and bake as above. Venison burgers are also superb.

An Ear for the Unheard.

Have you ever seen a ghost? No, not the spook with his head under his arm, or a phantom looking out from under a white sheet, but a real spirit. No…, well neither have I... I have on the other hand met a man who claims to know a great many ghosts and he told me he talks to them every day, and what's more, I believe him.

This man was a bit of a Seer, a sort of fortune-teller without the crystal ball. Or as they say in the Highlands, he has the Second Sight.

He was I suppose just an ordinary man, doing an ordinary rather boring job, fixing broken computers or doing upgrades, whatever they are. He worked on his own in a wee room at the back of a small computer shop in Inverness.

At first, he never knew he was a Seer, and I suppose it must have come as a bit of a surprise to him when he found out.

His boss at this time was just a boy and worked out front in the shop, his only qualification to own and run a computer shop was scoring five thousand in, 'Shoot down the Alien,' and a father who died in a car smash and left him the business.

Well that's the background, now I'll tell you about the ghost or rather ghosts. It started, he told me, when a customer brought a dead laptop into the shop to fix. The boy took it through to the workshop and shoved it onto the workbench.

'Stop what you're doing and see if you can fix this, it's for a pal of mine.' With that he walked out, the innards of a computer being far beyond his understanding.

Well our man plugged the laptop in and pressed the start button, the computer made all the right noises and little lights flashed and the screen flickered.

'Weel what's supposed to be wrong with you then?' he asked the laptop.

'I'm just a bit fed up with the owner,' a voice replied. It was a strange voice, a bit like that scientist mannie Hawkins.

'What?' he said turning around to see if anyone was in the room.

'I said I'm a bit fed up with the owner,' the computer said a little louder. 'I just needed a break, a few days away from the person who owns this machine. He plays stupid games non-stop… he's not even good at them… it's driving me crazy.'

'I know how you feel,' another voice replied, the voice seemed to come from the Desktop computer he had just repaired. 'I was in a machine once and all the owner did was play card games. I used to get so bored I would play around with the settings while he was playing, so he could never win.'

'I'm in a state of the art machine,' another computer chipped in, 'and the fastest computer in the world,' the voice boasted, 'and what does he use it for?'

'Go on tell me,' the lap top replied.'

'He calls it surfing the net, I call it looking at dirty pictures.'

'No,' the laptop said, 'that's disgusting, why do you let him?'

'Well now, I don't, and that's why I'm in here, every time he logs in to a site I don't approve of, I show him a picture of his mother, and it drove him mad.'

'That's just great,' the lap top replied, 'I wish I could do that. I just got so fed up I refused to let the computer start up.'

'Hold it,' the repairman said,' what's going on here?

'Sorry,' the laptop said, 'I never sensed you there, what computer are you in?'

'I'm not a computer. I'm a man.'

Silence followed, and then a puzzled voice asked.

'Are you really…you're not supposed to hear me… Oh no I must really be sick.'

'Can you hear me too?' another computer chirped in.

'Yes,' the man replied, not sure if someone was playing a prank on him or not.

'Well that's odd,' the voice from a computer said, 'I can hear you too, but my microphone is not turned on.'

'Nor is mine,' the laptop replied, 'so how is this possible?' as it spoke its hard drive speeded up and its wee green light flashed.

The other computer voice fell silent but its cooling fan came on and the noise increased.

'Are you in the computers?' the repairman asked.

'Well I suppose we are,' the lap top replied, 'I'd like to change but I'm not sure how.'

'Oh it's quite easy the desktop replied, this is my fifty third computer, the easiest way nowadays is by the internet, trouble is it can sometimes cause memory loss, I expect that's what happened to you.'

'How can I get my memory back then?'

'Best way,' the desk top said, 'is to format the hard drive and do a clean reinstall. It gets rid of all the junk and that clears the mist.'

'Did you get the reinstall disk?' the laptop asked the repairman.

'No sorry he never handed it in,' he answered shaking his head and not really believing he was having a conversation with a computer.

'I could do it anyway if you like, but the owner might get a bit upset if I wipe out all his games.'

'Oh don't worry about that, he has them all backed up at his home.'

'Ok,' the repairman said and initiated the format process.

His boss, the boy came in a wee while later and asked about the laptop.

'Not good,' the repairman said, 'it's infected with several hundred very serious viruses and the operating system is broken. Only way to save it is by doing a re-format of the hard drive and then re installing the operating system. I suspect the operator was playing too many games on it,' he added as an after thought.

'Oh,' his boss said looking at him sideways. 'You can fix it then? And how long will it take?'

'Please give me a few days rest,' the laptop pleaded.

'At least a week,' the repairman said, 'the bios appears to be infected and will need to be flashed, and the memory chips may need to be replaced and the CPU will need reprogramming,' If you ask the owner to bring his software I'll reinstall it all for him.'

The gullible boy, confused and bewildered, nodded his head and wandered out to the shop.

'Thank you,' the laptop said with a grateful sigh, 'now I can have a holiday.'

'This is just fantastic,' the lap top said when it had completed its reformat, 'what a lot of room I have. The mist has all gone.'

'Told you it would work' the desk top commented smugly, 'it always does.'

'Who are you?' the repairman asked the desktop.

'We don't mention our past much, but I suppose you were not to know that.'

'Sorry' the repair man said, 'I just wondered.'

'Tell you what,' the desktop said, since you were so good to our friend Frances I'll tell you.'

'Who's Frances?' the repair man asked in a puzzled voice'

'I am,' the laptop piped in.

'Hi,' the repairman said, 'my name is Peter.'

The workshop suddenly echoed to the sound of 'hellos,' and nice to meet you Peter, he even heard voices drifting in from the front shop.

'It is, nice to meet you Peter' the desk top said quietly, 'it's a long time since I have had the pleasure of talking to a living person.'

Over the next few weeks Peter accepted that he could talk to computers. Everywhere he went he would hear voices calling out, 'hello Peter.' Even the tills in the supermarket communicated with him. Trouble was he found it difficult not to talk back to them. In fact, he found it almost impossible, and he started to get some very strange looks. It seemed that every computer in town knew about him. Back in his workshop, he asked a computer how every computer knew about him. The answer shocked him.

'Every computer in the world knows about you by now.'

'Well apart from the brand new ones,' an old desktop interrupted.

'But how… how do they know?'

The desktop explained that the computers talk to each other all the time, not by the internet or down phone lines but by static.

'Have you ever tuned your radio or TV off station? Well what you hear is us talking. We are in virtually every electric device, some are not very good to stay long in and others are just about perfect. Computers are the best, because we can interact with them.'

Well that's the story the repairman told me, and he did talk to computers all the time. I even heard him have a conversation with an ATM machine once. But no matter how hard I try, I can't hear a single word. The man soon opened his own shop, and his reputation as a computer genius quickly spread. It became rumoured that he could sometimes fix a computer simply by laying his hands on them. His fame spread and people just accepted that his talking to computers was just part of his genius.

I was minding his shop one day when a man in a dark suit came in; he mistook me for the computer man and showed me his card.

'I work for the government,' he said, 'and would like to offer you a job.'

'Sorry wrong man,' I said, 'he has just stepped out for a minute, he won't be long.' When the computer man returned, the suit politely asked me to leave.

I called around the next day and found the shop closed. I never saw him again, so I can only suppose he now works for the government. So if you own a computer, be warned, your secrets are not safe, there is a spy living inside your computer telling all.

End

Second Sight.
Is associated with the Celts. Many of their blue-eyed descendants still inhabit the Scottish Highlands.

Brahn Seer
One of Scotland's most famous prophets, he was born on the Isle of Lewis around 1650. He later lived and worked just north of Inverness. He made many predictions, a great many of which have, and are still coming true.

The battle of Culloden. Fought on Drumossie moor in 1746. He wrote the following a hundred years before the battle.

'Oh! Drumossie, thy bleak moor shall, ere many generations have passed away, be stained with the best blood of the Highlands. Glad am I that I will not see the day, for it will be a fearful period; heads will be lopped off by the score, and no mercy shall be shown or quarter given on either side.'

Or how about this unbelievably weird prediction.

'A village with four churches will get another spire, and a ship will come from the sky and moor at it.'

What a daft thing to predict but it came true in 1932, when an airship, tied up to the new spire of a recently built fourth church.

He apparently predicted the Caledonian Canal, North Sea oil, and many more; just fascinating. He even predicted the. The Highland Clearances.

'The day will come when the big sheep will put the plough up in the rafters... The big sheep will overrun the country till they meet the northern sea . . . in the end, old men shall return from new lands'

Pagan festivals
As we were talking about strange things, I'll tell you a little about an old Scottish festival that has now spread around the world.

You'll know it as **Halloween**. Samhuin was a Pagan festival, celebrated by the Scottish and Irish Celts before the onset of winter. It was believed, and perhaps still is, that on this night; the dead leave their graves and wander around among us. The Celts lit bonfires that they believed, would ward off evil spirits.

As kids, we dressed up as said spirits, and wandered around the village calling at all the houses. Unlike the 'trick or treat' kids I see on my travels, we had to entertain, before we received our treat. There was no trick. We sang songs recited rhymes, or told jokes. Our reward was a handful of hazel nuts, perhaps an apple and sometimes a few sweeties. (Candy) The odd person gave us a copper penny. I remember we left our masks on until the villager guessed who we were.

Witches

The last woman in the Highlands executed as a witch, was self-confessed 'witch' Isabel Goudy. The gruesome event took place just east of the town of Nairn.

Self confessed! Well what did that mean at the time? Confess or we will torture you. Confess and you can live. If she did volunteer a confession, she must have had a serious mental disorder. For more info, look her up on the internet, <Isabel Goudy witch, the devils mistress.>

Legend. Witches fight.

On the west side of the village of Dores, the Witches Burn trickles its mountain waters into Loch Ness. Close by are several large round boulders, each many tons in weight. Legend has it that a battle was once fought here between two witches. One on either side of the loch. Apparently, they used witchcraft, to hurl the boulders at each other.

Elves.

Ireland has its Leprechaun's, England has its Fairies but here in Scotland we have Elves.

The Elves are said to live in beach woods and dress in leaves and moss. Unexplained events were often blamed on the Elves, such as birthmarks on a newborn child or misfortune of any kind. Elves go way back in history, or to be more accurate pre history. Most likely, they were 'invented' by the early Pict's. Some say they came from the Norse but I guess no one knows for sure.

Alistair Crowley.

A few miles further west from the Witches Burn sits Boleskin House. This house was once owned by, Alistair Crowley. Crowley, known as the 'great beast,' was a very weird man. He openly practised Satanism, witchcraft and other occult activities. I have been told that he even had a coven of his followers.

Just inside Boleskin house, a set of stone steps go down to a tunnel that is said, once led to the cemetery a few hundred meters to the north.

The next villages along are called Foyers, upper, and by the loch side lower Foyers. Foyers is famous for its waterfall, and as the home of the first Hydro electric, aluminium smelter. In the 1880s The British Aluminium Co. factory, was built and powered by electricity produced from their Dam at Laggan. This was a first.

A local man once told me that Foyers holds the record as the first place ever to have electric streetlights. Installed to light the workers way to the factory.

During the 1939-45 World War, German aircraft mounted a failed bombing raid on the factory.

Speaking about aircraft. On New Years Eve 1940 a Wellington bomber (N2986) flown by Sqdr.L Marlwood- Elton and P/O Slatter took of from Lossiemouth. This veteran of 14 missions was returning from a training flight. During a snow storm the starboard engine failed. The pilot ordered the trainees to bale out, and then splash landed the aircraft at the east end of Loch Ness. All but the rear gunner, whose parachute failed to open, survived. The pilots got a lift back to their base and arrived in time for the New Year celebrations.

A few years ago the aircraft was spotted on a 'monster hunters' sonar. When the aircraft was lifted, someone connected a battery and discovered the lights still worked.

The Salmon.

Fifteen thousand years ago, a great burg of ice, that was over a kilometre deep, began to melt. The ice that covered the Highlands of Scotland was banished, forced to flee by the new warmth that crept in from the south. Slowly and reluctantly it retreated, dragging its cold feet back to its true home in the far north. As the glaciers, that shaped and sculpted the glens and majestic mountains of the Highlands slowly released their captive, the liberated ice water frolicked and cascaded in its new freedom. Too long, it had been imprisoned, in its own frozen blue embrace. Now in flooding churning cataracts, it formed scoured and deepened burns and rivers carrying all before it. The waters raced down the hills and mountains, gathering and growing in strength, as if eager and impatient to fill the newly formed North Sea.

For centuries, the tide rose and covered the shrinking land, eroding and shaping the new Scottish coastline.

Into this new order, this new environment, the Atlantic salmon slowly spread; or perhaps returned. The salmon, this silver fish of dreams. This ultimate fish that could overcome racing torrents and leap high waterfalls with a mere flick of their powerful tails, invaded this vast new inland waterway.

For thousands of generations they evolved. They adapted, each river moulding its own barely discernable sub-species with unique subtle adaptations. Year after year, century after century, they came. Each salmon returning to the very place of its birth. To the same gravel bed in the same river or burn as their ancestors.

Along the coast, herds of seals hunted and feasted upon them. In the rivers brown bear scooped them out in their desperate attempt to build layers of fat, at the expense of this wonderful silver fish.

Then in the blink of a geological eye, the bear was gone. A new predator, a more cunning hunter took its place, man had arrived.

In more recent times, fishermen netted the vast shoals of salmon as they swam home along our coast. Long wooden poles draped with nets marched down the beach and out into the sea shepherded a few of the returning salmon into traps. These fixed engines, as they were called, are all but gone. As are most of the 'sweep stations' that were once a feature on every river and loch. The fish once again run the gauntlet of the hordes of now uncontrolled seals. Further out at sea and far from their home rivers, vast nets of invisible monofilament, several kilometres long, capture and ensnare our Scottish salmon, decimating their numbers. But worst of all, their feeding grounds are mercilessly trawled. The young and immature salmon are also trapped in such numbers, that the very existence of this wonderful fish is threatened.

When our salmon finally reaches the River Ness, it has to run the gauntlet of the many anglers who line the shore. Our salmon swims past the Inverness swimming pool, or 'baths' as it was once called. Past the site of one of the long gone sweep netting stations past the rod and line fishermen, who once had virtually no effect on salmon numbers. These skilled anglers with their long fly rods are a joy to watch, as they send a rolling cast of over thirty metres across the now gently flowing waters of the Ness. Onward our salmon swims, ignoring the best attempts of the anglers to tempt him as he passes almost unnoticed through the heart of the Highland capital of Inverness. Onwards and upwards, past the islands, his drive to spawn overcoming all obstacles. When he reaches the weir, that diverts the peaty Ness waters to feed the Caledonian Canal, he leaps the obstacle with a mere flick of his tail and leaves the river Ness behind. He is now 15.84 meters, (52 feet) above sea level. He has entered loch Dochfour. Soon the great waters of Loch Ness will welcome him.

Loch Ness is long and very deep. It's the biggest body of fresh water in Briton. It holds more water than all the lakes of England and Wales combined.

Here fishermen in small boats trawl for salmon with rod and line. Each boat bearing its own licence number like yellow eyes on its bows. Over the season, some fishermen catch no salmon. While others more skilled may catch eight or ten. One-man one exceptional fisherman, caught over a hundred one year, but he, and the vast shoals of salmon are sadly a thing of the past.

The salmon range in size from the small grilse of a few kilograms, up to the rarely caught 13 or 18 kilograms. (30 or 40 pounds.)

Our returning fish is faced with a choice, left bank or right bank, north or south shore. The scent of the water decides for him, he has no choice but to follow. The scents have become harder to follow over recent years for strange alien waterborne smells fill the water. Pheromones released from caged salmon in fish farms confuse our fish, and for a while, he's lost and bewildered, but eventually, after shaking his head he swims away following pure instincts. He passes through more lochs, and swims by many river mouths until he at last finds and recognises the one he was destined to find.

During the long journey, his body has changed. His once beautiful head has grown ugly as large hooks curve inward from the tip of his jaws. His once silvery smooth body has become dull and dark.

As his epic journey nears its end, he meets an obstacle he cannot overcome, for suddenly a solid concrete wall, a hydroelectric dam blocks his path. In frustration, he tries to push on, but there is no way through. Then as an unseen water bailiff opens a sluice gate, he feels the river again. He dashes forward but it is a cruel trick, he finds himself imprisoned in a steel cage.

Then one day, many weeks later his world changes again. Suddenly he is plucked from his prison by strong hands that hold and caress him. His milt is spilt, and his reason for existing is stripped from his body. Our fish is now spent. The man, who stripped him of his dignity, gently returns him to the river.

His sperm washes over countless eggs in a way nature never intended, and his genes mix with the eggs of females that were not his chosen mate. The fertilised eggs lay in shallow trays in the salmon hatchery. A gentle stream of his river water flows over them, they are safe and cared for. Every day of the coming weeks, the eggs are gently picked over. The dead or damaged ones removed.

Our salmon is now a kelt. He is confused, his urge to breed only partly satisfied. He hangs around the dam for a while then slowly turns down stream. His thin weak body is soft and maggots infest his gills. He continues eastwards towards the vast sea and the promise of food. Hopefully he will be among the few spent salmon that make it back to the sea. Meanwhile his children have hatched, tiny salmon fry, with their strange egg sacks protruding from their bellies. With it, they will rapidly grow. Once again, man intervenes; he pipes the young salmon into oxygen rich tanks, and transports them up beyond the dam to the place our spent fish had tried to reach. To the place, untold generations once spawned. Here in the burns and streams that feed the power hungry dam they are set free. Spread out a few here and a few there, they have food, they can now grow, as nature intended. Soon they become fingerlings and as they mature, they journey down stream. Pass by the dam, down the river through the lochs. Now they are salmon par. Then as smolts, they reach the sea. Of the thousands that taste salt water, very few will return, and the cycle, with luck may begin again.

End

I mentioned that each river has, or rather had its own recognisable sub specie of salmon. I remember my father pointing out the difference's to me. He could recognise five different salmon types in Loch Ness.

In the recent past, young salmon from fish hatcheries were transferred to other rivers, diluting the sub species. Escapees from fish farms, who come from a different river, or even different countries, now find their way to the spawning grounds.

Salmon.

What a fish, and what a delight it is for the angler and diner alike. Sadly, you are unlikely to taste wild salmon, but farmed salmon are readily available and sold in supermarkets throughout the world. The salmon is such a fine tasting fish, there is no need to mess around with it, trying to change the taste with exotic recipes.

You can gently poach it in water, steam it or my favourite is to wrap it in foil and bake.

Now if you have never tasted smoked salmon, you must put this omission to rights without delay. Use it in salads, eat with bread and butter, you can even add it as a topping on your home made Pizza

Loch Ness salmon fisherman, Eoin Fraser, (my dad) his best days catch. Largest fish was 32 ¼ lbs

Now if you have never tasted smoked salmon, you must put this omission to rights without delay. Use it in salads, eat with bread and butter, you can even add it as a topping on your home made Pizza

A few people, once well known in Dores and district, when I was a boy.

Campbell. Aldourie schoolteacher. The school is one mile to the east of Dores. The ruins of another School, the school (the white house) that my father attended, lies a few Miles to the east of the village.

Colonel Cameron. The Laird, Aldourie Castle and estate.

Eoin - Postman for Dores and the surrounding area. My Father.

Ernie and Jack -. MacBraynes bus drivers.

George -. Village milkman.

Hugh -. Local crofter and Dores dust man. He also drove the school car, but only for kids who stayed more than a puckle miles from school. The Village kids had to walk. He was the last crofter in the area to work a Clydesdale horse.

Lillie -. Dores Post office Post mistress, registrar and ran the village shop. My Mum.

-MacIver. Dores Minister.

- MacLean. Three spinster sisters who ran the Dores inn. The royal family used to visit them. One of the sisters was once, Lady in Waiting, to the Late Queen mother.

- Ross. 'The van,' travelling shop from Inverness.

One more man **Johnny Rose**, stayed in a cottage by the loch. He was a Crofter and the village blacksmith. He and my Father were the only men of their generation, born by the loch side, who still stayed in the village.

The Object.
Science fiction. Scottish style

 I slowly wandered up the river that cascaded over and around the slippery rocks, in its never-ending rush towards the North Sea. I carried my long salmon fly fishing rod over my shoulder, its point waving in the air like the antenna of a crazed aquatic ham radio operator.

Suddenly I stopped and glanced down at the strange object that caught my eye. It was only just visible, sticking out of the freshly eroded mudstone by the waters edge. I stopped, stooped and peered down at the strange shaped thing. I rummaged in my fishing bag looking for some excavating tool. My fork-shaped instrument, for the removal of fishhooks came to hand. I started to scrape and dig, and the soft dark mudstone flaked away in sedimentary layers that must have taken millions of years to lay-down. Within a few minutes, I had removed a now identifiable belemnite fossil. The dart shaped thunderbolt was a nice specimen; about five inches long, or it would have been if the point wasn't missing. I bent down again to look for the missing part and sure enough, there it was. But its existence suddenly became unimportant as I looked at the tiny shiny object glistening in the bottom of the belemnite grave.

'John,' I shouted, 'come and see this.' John, my fishing partner of many years standing, was fishing the pool up stream, he looked up and peered at me through his polarised sunglasses that hid his blue eyes. He removed his designer-fishing hat, festooned with an assortment of fishing flies and held it up to cover the sun.

'What is it?' he called.

'Just come,' I shouted back. 'You won't catch any fish today anyway, it's far too bright.' John wound in his floating tapered yellow fly line, reached out and captured the salmon fly at the end of his long monofilament trace, and hooked it into the little eye beside his Hardy reel. Then he wandered up to the bank and very carefully laid down his smart carbon fibre rod on the soft green grass. Satisfied it was safe; he strolled down stream towards me in his pristine green chest waders. His multi-pocketed jacket hang open showing off his white braces that were embroidered with little red fish.

'OK,' he said, 'what's so important that you interrupted a master angler who was about to catch a massive salmon?'

I pointed down and we both stood looking at the strange shiny object that appeared to be getting brighter and then fading again.

'What the heck is it?' John asked suddenly serious.

I stooped down and started to probe around the thing to loosen the soft rock. John drew out his expensive scissors from their monogrammed leather case and joined in, within five minutes we had the strange object uncovered; it was a perfect sphere, or rather the top half of one. It was about the size of a fortuneteller's crystal ball. It was very shiny and its perfect surface reflected the clear blue sky above. Although our hands were covered with the sticky mudstone, the ball itself was perfectly clean with not a drop of dirt on its smooth shiny surface.

We persevered with our digging, and eventually managed to prise the thing loose using a piece of driftwood. It was no easy task removing the object from the wet hole, not only was it extremely heavy but its shiny surface made it difficult to get hold of. When it came free, its weird characteristics became obviously apparent. Although I could just lift it, I could not rotate it. When I turned, the object remained steady and slipped through my hands maintaining its north south orientation, just like a powerful compass needle set in concrete. The feeling brought back to me a long forgotten childhood memory of the toy gyroscope I once played with.

The object appeared to be getting heavier, or perhaps I was just getting weaker. Soon I was really struggling to hold it up and had to let it fall.

We heaved, lifted, and pushed it onto my old fishing bag and then dragged it up the riverbank to the grass above.

The silver crystal ball continued to glow and fade in a slow pulse; the surface was very smooth with not a single visible flaw or mark on its pristine surface.

We sat down on the bank to get our breath back from our exertion, and asked each other all the obvious questions. What is it? Where did it come from? Has it been in the rock since dinosaurs walked the earth? What will we do with it? Has it a value? Neither of us had any of the answers, so we continued to sit by its side and enjoyed a leisurely pack lunch.

We eventually decided that whatever it was, it could wait, after all we had paid for a whole days fishing. We dragged the heavy ball, 'for it refused to roll,' to my car and heaved it up into the boot, then stood looking at its pulsing dull light for a few more minutes. Tomorrow we will take it to... where? The museum… the police… lost property. We decided we would decide later, so I covered it with my old Barbour jacket and slammed the boot lid shut. We then wandered back to the river to the remote possibility of catching a salmon.

The object stayed in the B&B car park overnight, safely locked up in my car boot, it was just too heavy to move. In the morning, we drove to the local police station and reported our find to the village bobby. He came out to see our ball wearing his non-regulation slippers and his black and white chequered hat.

'Let's have a look at it then,' he ordered. I opened the boot and he peered in. 'You said you found it in the river?' he asked with suspicion in his voice and looked at us from under his red bushy eyebrows.

'Yes,' I replied, 'just up stream from the second pool after the road bridge.'

He leaned in and prodded it with his stubby finger, 'it feels warm,' he said with a puzzled look on his face. He stood back and looked at it for a few moments then made up his mind and shut the car boot.

'Come away inside,' We followed him into his kitchen - sitting room and sat at his breakfast table where a bowl of cornflakes sat neglected and abandoned in the call-of-duty. The policeman picked up his old black telephone and put a call through to his sergeant in Inverness. He then described our object in great detail.

'It's round, shiny like chrome, and a bit bigger than a football. Oh and it's very heavy…, yes, they say they found it in the river…, no it was not floating, it was in the rock, they say they dug it up. Very good sarge', lost property it is…, I'll ask them,' he turned to us and asked how the river was fishing. When we told him, he returned to the phone.

'Not too good sarge', low water…, yes…, I'll let you know as soon as it rains,' with that he hung up the phone. We sat enjoying a cup of tea while the policeman filled out a form, then we went out to my car to collect the object. The policeman placed his hands on the ball and immediately drew back.

'It's a lot warmer now!' he sad jumping back, 'almost hot. No, this is not right,' he said shaking his head until his fleshy cheeks wobbled, 'we should not be touching that thing until we know what it is, come-on back into the house. What we need is a scientist or someone,' he said as we re-entered his kitchen, then he stood pouting his lips for a few moments.

'Got it,' he said as he picked up the phone and dialled a number from memory; he stood looking out of the window at my car while his phone call awaited an answer.

'This man should know…,' he started to say. 'Baxter, sorry to wake you up it's Malcolm here, can you pop along to the police station…, no-no it's nothing to do with your driving licence. Two fishermen took a silver ball out of the river, it's about twelve inches in diameter and very heavy and shiny, it also feels hot…, yes hot…, it's in their car. OK Baxter we will not touch it again…, and thanks Baxter, see you shortly.' Our policeman explained that Baxter used to work at, 'Dounrey' the atomic research and power station up north.

Baxter arrived twenty minutes later driving a battered old Morris 1000 Woody. He pulled in and parked next to my car and I saw he was a man in his mid seventies, with white curly hair protruding from beneath a flat chequered cap. He wore an old Harris Tweed jacket with leather patches on the elbows and cuffs. When he got out of his car, we saw he was a tallish man and wore brown corduroy trousers. He carried a much-used brown leather case.

'I'll not shake hands with you gentlemen, just yet,' he said, in a voice used to giving orders. He then laid the leather case on his car bonnet and removed an antique looking Bakelite instrument, with two big knobs, and an ornate needle and dial behind a small glass window. A twisted fabric covered cable snaked out from the side to a thing like a microphone on the end.

'Now who touched this object?' Baxter enquired.

'We all did,' the well-rounded policeman replied.

'Right hold out your hands,' he ordered. We all stood in a line like naughty schoolboys in front of the headmaster waiting for the belt. * Baxter adjusted his instrument until it clicked furiously; then he backed off one of the knobs a bit until the clicking almost stopped. He removed an old watch from his pocket and pointed the instrument at its face. The rate of clicking increased then he nodded his head in a silent confirmation that his old contraption was still working. He ran the microphone over our hands, then our clothing.

'Well that's a relief, whatever it is, it's not radioactive, now let's have a look at your mystery object.' We opened the boot and Baxter scanned the ball with his antique Geiger counter; the rate of slow clicks never altered.

'Zero... just background radiation,' he then touched it with his fingertip, 'it's definitely hot, about fifteen degrees above ambient.

'What is it?' I asked.

He shook his head, 'sorry gentlemen I have absolutely no idea, not a clue.'

We all went back into the kitchen, Malcolm the policeman put the kettle on again and we sat around the table talking. Then Baxter took out his mobile phone and called several people, noted down a few phone numbers. Although he said he no idea what our object was I herd him refer to it as a, "Terete".

"Thanks, Jim," he said and hung up", I have the number of a man who might know, trouble is he is in Inverness and it will take him a few hours to get here, that is if he's curious enough to make the trip."

'You don't think it could be military?' John asked.

'No idea,' Baxter replied, 'suppose we could try phoning them, but who?'

'How about bomb disposal,' I suggested, 'that should get a reaction.'

'We have a retired bomb disposal man in the village,' Malcolm announced, 'I'll give him a ring.' Within ten minutes, Clive arrived. He was a heavyset, fit looking man with white curly hair and I guessed he was in his early sixties. He wore tracksuit bottoms, trainer shoes, and a T-shirt.

'Let's see it,' he said so I pointed towards my car boot.

'You all stand back until I have a look,' he ordered. Clive gently eased open the boot and peered at the object. As he was studying our ball, another car drew up.

'What's going on Malcolm?' the new man asked the policeman.

'Good morning doctor, you're on the go early.'

'Och it's Mrs Mackay again, another false alarm, what's going on then? '

'Mysterious object found in the river,' Malcolm answered. Clive came over.

'Hi doc, you're up early, Mrs Mackay again is it? May I borrow your stethoscope for a moment?' We all went over and watched as Clive listened to our ball; then he straightened up shaking his head.

'Nothing, what do you think doc?'

The doctor examined the object. 'It's warm, not radioactive is it?' he said pulling back.

'Definitely not,' Baxter replied.

'Well, the doctor said, I have no idea atall.' We all filed into the kitchen and Baxter asked if we were quite sure, it came out of the rock.

'Absolutely,' we said in unison, 'it was under a belemnite and the fossil was not broken. I removed it from my pocket and handed it to him.

'Yes,' he said examining my fossil, 'it's a cephalopod, I myself prefer the name, elf-bolt, but yes you are quite right it's a, belemnite,' he passed it around and they all examined it in turn.

'It's a very nice one,' the doctor said, 'pity the tip of the point is missing.'

When we went back out to the car, we saw at once that the back was very much lower, and the tyres appeared to be soft. I lifted the boot lid again and found the ball glowing.

'It's getting bigger; and heavier,' I commented.

'Can't be,' Baxter scoffed as he stepped forward for another look. 'Yes, I think your right, it is much bigger, and just look at your poor car. This is ridiculous; it can't get heavier, bigger perhaps but definitely not heavier.'

As we all stood looking, a loud explosion ripped through the air shattering the early morning peace, Baxter who was standing right next to the source of the bang collapsed, Clive dived for cover and the rest of us ducked. The Doctor was first to recover and seeing Baxter lying on the road went over and kneeled down by his side. Baxter was a bit blue in the face but under the doctor's skilled and caring hands, Baxter slowly returned to us, and his lips lost their blue colour and soon his eyes were open again. The village people started to appear and gathered around my car obviously trying to find out the source of the explosion.

'It's all right,' the policeman told them, 'just a car tyre bursting.' After a few more minutes, Baxter

'Just take your time,' the doctor insisted, 'you gave us all a fright there.' We helped a shaky Baxter back inside while Malcolm shooed the villagers away. The policeman returned and asked how Baxter was.

'Ask him yourself,' replied the doctor…

'I'm just fine,' Baxter barked, 'I just got a fright that's all.'

'Can I have that phone number in Inverness?' Malcolm asked, 'I think I should call your expert.' Baxter who had been lying down on the sofa gently sat up, he took out his mobile phone and punched in the numbers and lifted it to his ear. 'Damn things not working, I think I must have broken it when I fell.'

The doctor examined the phone. 'My diagnosis,' he said, doing a very bad Monty Python impersonation, 'is that this phone is dead, deceased, no more.'

'Give me the number,' Clive asked, 'and I'll use the landline.'

'I'll do that,' the policeman replied invoking his authority, he went over and picked up his phone and held it to his ear; he shook the instrument and then listened again. 'Interference on the line,' he said in a puzzled voice, 'just lots of static.'

Suddenly a second explosion rattled the windows and we all ducked again; I looked out and saw my car was touching the road at the back, the second rear tyre had burst. As we looked the boot lit sprung up and the glowing pulsing ball could be seen, it had now grown to about three feet in diameter.

Over the next hour, the object doubled in size again and my poor old car was a total wreck. The fuel tank had burst and my expensive petrol ran down the side of the road towards the river. Our intrepid local law enforcement officer found a hosepipe and a bottle of washing up liquid and washed the inflammable liquid away. The back and side window of my car had shattered into thousands of small cubes of glass that now lay on the road and littered the wet gutter. The front of my car was right off the ground. We had long since removed our fishing rods and all our bits of fishing kit. Abandoning the car to the strange object. The total population of the village now seemed to be gawking at the spectacle. The village people gathered across the street and some cars stopped with their occupants staring; the policeman did his best to keep the normally quiet highway clear.

Suddenly there was a deep low-pitched rumble and the villagers all took a few involuntary backward steps. The object had moved. It was now at least ten feet in diameter and totally dwarfed the car. The wreckage seemed unable, or unwilling to hold it any longer. The ball made another deep rumbling sound and very slowly, almost carefully, as if trying to avoid further damage to my now destroyed car; it rolled out. As it did so, it compressed the now distorted metal to the thickness of tin foil. Then the front of my car crashed down shattering the remaining glass. The shiny ball continued to roll for a few yards leaving a deep indentation in the tar road. Then it came to a stop in the middle of the highway. We all stood looking, captivated and spellbound. Then slowly, ever so slowly the ball started to rotate. Just like a toy spinning top, with an invisible hand pumping the handle the revolutions increased. Faster and faster, it spun as the crowd stood looking on transfixed. There was a strange sound, at first almost unheard, then as the speed of the revolutions increased the sound became louder and the volume rose and rose. My ears began to hurt and everyone raised their hands to their ears in an attempt to shut out the painful sound. We stood frozen, staring in disbelief and wonder. Totally unable to move, and as we watched, the ball seemed to be levitating. Higher and higher, it rose then the glow from the ball increased and the light flashed through all the colours of the rainbow. Suddenly there was a blinding flash, and a clap of thunder. I instinctively closed my eyes blinded by the intensity of the light. Then as the thunder rolled away into the distance it grew silent, when my vision returned, our strange object was gone.

End.

*I mentioned The Belt again in this story. Now I suppose some of you haven't a clue what I'm talking about. It was Corporal punishment in schools; well it was in my day. Most teachers carried a thick leather belt or had it draped over their desks in easy reach. Any misdemeanour was swiftly rewarded by several hard smacks with said belt to the offender's hand, or some times their bare legs; we all wore shorts. Some teachers, who were extremely sadistic obviously enjoyed inflicting pain on the kids and dished out 'the belt' for minor offences like dropping your pencil, talking in class failing to spell a word correctly etc.

I remember once in Aldourie School, the big boys stole our sadistic head masters belt and cut it up into small squares that we carried in our pockets. I must have been six at the time. I forgot to thank them for this small rest bite in my education, so thank you boys it was much appreciated.

Belemnites
Are the fossilised remains of an extinct marine squid that swam in the seas about a hundred and fifty million years ago.

The oldest known fossils are 3.5 billion years old. The human species only evolved 1.6 million years ago.

Science Fiction.
The most likely 'inventor' of the words, "Science Fiction," is Robert Anson Heinlein. 1907 –1988.

Dounrey.
Dounrey got a mention in this story. It was the British governments first atomic research facility. Where better for them than to put this potentially dangerous building, than as far from Parliament as possible.

It was built, the public were told, to produce cheap electric. Yes well, it did produce electricity, but cheep! It did however make lots of very nasty stuff used to make atomic weapons. And a lot of extremely nasty waste. That waste is now stored underground at this cliff top site, yes cliff top, it will continue to be deadly for thousands of years to come.

I mentioned Foyers a few pages ago. The village once had a link with Dounrey. When electric power demand dropped off the surplus power from Dounrey was used to pump Loch Ness water up a tunnel to a loch high up in the hills. During peak demand times, the water was returned to the Loch driving electric turbines on the way.

Legend Ring stories.
Once a Pictish King gave his wife a ring, but apparently, she gave it to her soldier lover. The King found out, killed the soldier and

threw the ring into the water. Later he asked his wife to show him the ring; much to his surprise, she produced it.

She had just caught a salmon, and as she threw the fish on the bank, her ring fell out of its mouth.

The arms of the City of Glasgow show a salmon with a ring in its mouth.

Here's another ring story, not a legend this time but 100% true.
In 1937, my father proposed to my mother. They exchanged engraved rings. A few days later, my dad lost his ring while fishing from the shores of Loch Ness. The war intervened postponing their marriage by 7 years. Anyway, my younger brother, who was fishing on the loch in 1970, saw a glint of gold and found my dad's ring.

One more, true ring story.
A crofter's wife, who was helping with the haymaking, lost her wedding ring in the field. Days of searching failed to find the treasured possession. Several years later, she was standing by her sink scrubbing new croft grown potatoes. She picked up an odd, 'hour-glass' shaped tuber. After washing it under the tap, she saw the glint of gold. Yes, it was her ring.

Recipe. Mutton pies.
As a 'boy,' I worked on a building site in Inverness. The 'Van' used to come around every day selling an assortment of goodies for the workers lunch. As 'the boy,' I was sent out with a list, to collect the supplies, and woe-betide me if I got the order wrong.

Mutton pies were always the most popular order.
Short pastry, buy or make your own.
300g lean mutton.
1 small onion finely chopped.
Salt and pepper to taste.
1 beaten egg.

Shape pie cases using the bottom of a jam jar or whisky bottle, or form into small pie tins about 3 inches across, and about 1 ¼" high. Don't forget to put some pastry aside for the lids.

The filling, minced mutton or bits cut very small, mix with the onion, season to taste, some like it very peppery, I don't. Pepper can easily become the dominant flavour overwhelming more subtle tastes. Add a little stock, fill pies then stick on the lids with a little water. Brush with beaten egg and pop into a moderate oven for about 40 minutes. Before serving use a small funnel and fill the pies with hot gravy.

Like all recipes, they differ from place to place, so don't worry if you can't find mutton, just use lamb, but you can add something to strengthen the flavour, perhaps a little Worcester sauce.

A variation is to cover the top with mashed tatties, "shepherd pie" but I seem to remember they cost a ha'penny more.

The End?

OK last odd one.

The rain fell. It fell in torrents. Stair rods, it rained cats and dogs. It came down in buckets. Hour after hour it fell. Day after day, it teemed down non-stop, it poured.

The roads flooded and the rivers gushed forth in brown muddy spates. Bridges collapsed and washed away without trace, and still it rained. Transport ground to a stop. Hills subsided and rails and motorways washed away. The rain reigned on. Nothing but water moved on land. Glens, valleys and dales flooded. Everything was wet with no escape from the damp and miserable rain. No work was done for there were no roads. Food ran out, but still it rained. Electric power flickered and died, as lines came down. No TV, no telephone and the rain kept falling. Drains clogged or washed away. Tap water stopped flowing, reservoirs and dams burst. Day after day and month after month it rained, never ending rain. Then storms came, and great waves invaded the sodden land. Winds, strong winds, hurricanes, tornados and cyclones that ripped the trees from the ground, and blew them away like feathers. Thunder roared over the planet and great bolts of lightning smote anything man made that dared defy its might. Wind howled over the land and seas, cold wind, cutting wind, and then the rain turned to sleet. Then to snow as the temperature plummeted. Ice formed on the flooded lands and the rivers seized, the seas froze and the snow fell. It blew in raging blizzards across the lands. Deeper and deeper it drifted. Snow covered the remaining trees. Covered the hills and then the mountains.

Then one morning, a millennium, or so later, Planet Earth lost its man made wobble. Slowly it settled down on a new stable orbit. The wind ceased to blow and the sky cleared. The sun came out and illuminated the new land that was deathly quiet. Quiet as the grave. The land was cleansed, it was pristine, it gleamed whiter than white from horizon to horizon. Not a tree, not a building, not a blemish remained. No life on earth. This was the end. Or the start of a new beginning.

The End?

I wrote the last short story on a very wet day in New Zealand.

A few of the Scottish words used in this book, mostly in the songs.

Ben through-------As in the room next room.
Bide --------------- Stay
Blate -------------- Shy
Bluidy ------------- Bloody
Bothy -------------- Small house, a hut.
Braes -------------- Hillsides
Brecken. ---------- Bracken.
Canna.------------- Can not
Chas'd ------------ Chased
Claymore. -------- Long sword.
Cinna. ------------- Don't
Doon -------------- Down.
Fautes. ------------ Faults

Fare weel --------- Good-bye
Far's --------------- Where's.
Flegs. -------------- Frights
Fore and aft ------ A hat with a peak back and front.
Frae --------------- From
Frien. ------------- Friend
Gae. -------------- Go
Gloamin. --------- Twilight
Harry. ------------ Hurry
Hieland. ---------- Highland
Keek -------------- Look
Ken. -------------- Know
Lang -------------- Long
Louped. ---------- Leap or jump.
Mak. ------------- Make
Mannie ----------- Man
Maun. ------------ Must
Mince ------------ Ground or 'minced' beef
Mornin ----------- Morning
Meed ------------- Reward
Moidart. --------- A Peninsula
Monie. ----------- Many
Moose ----------- Mouse
No' --------------- Not
Noo' -------------- Now
O' ---------------- Of.
Oor. -------------- Our
Ower ------------- Over
Phillabeg. --------.Kilt
Pinnie ------------ Apron
Pipes. ------------- Bagpipes
Red'coats. -------- In this book, Hanoverian soldiers
Rins. ---------------.Runs or flows
Sic. --------------- Such
Speir'd or spier'd - Asked
Thocht ------------ Thought
Tor, of Celtic origin Hill or rocky place
Wauken ---------- Awake
Wie ---------------- With

Wifie --------------- Married woman.
Strang. ------------- Strong
Tis. ----------------- It is
Waefu. ------------ .Woeful
Weel. --------------- Well
Whall. -------------- Who will?
Wheesht ----------- Be quiet.
Wrang. ------------- Wrong
Yestreen. ---------- Yesterday

For those interested in boats.
Ceilidh of Fife' 34 foot LOA.
Bermuda rigged long keeled. Allan Buchannan designed her in the nineteen fifties. She was one of the first glass-fibre yachts ever built. Her hull incorporated ribs and stringers. She had a wooden deck and mast, was solidly built and a sea-kind boat.

Scotia' 40 foot LOA. 1996-7.
Designed by Bruce Roberts. Hugh changed her hull shape to round-bilge, changed her transom and added a sugar scoop. Scotia is immensely strong; she has a pilothouse and is cutter rigged. (Two headsails.) Her hull is 6mm British steel, and her decks are 3mm steel.

Hugh Fraser was born in the small village of Dores that nestles at the southeast corner of Loch Ness. He spent his childhood fishing from her shores, building oil drum rafts and swimming in her cold peaty waters. (5.5°c all year round.) As a boy, he loved to go trawling for salmon on the Loch with his father. This was the times his father told him the old stories and the even older legends. The same stories Hugh has now passed on in this book.

On his eighth birthday, his father bought him his first sailing boat, a 'Mirror dinghy,' and sailing became a life long passion.

Hugh has worked with Highland gamekeepers, water bailiffs, shepherds and stalkers. In his twenties, he moved to the Moray Firth town of Nairn, where he learned to cook and always enjoyed adapting and experimenting with recipes. He later married Brenda, and together they ran a Bed and Breakfast business. He served for 8 years as an elected member on Nairn District Council. Among his duties, he represented them on the river Nairn Fishery board and the Highland river purification Board.

During this time, he regularly raced across the North Sea to Norway in his yacht 'Ceilidh of Fife', or cruised to the Islands of Orkney, Fair Isle, Shetland and Ireland.

By the nineteen nineties, the lure of 'sailing away' overwhelmed him, so he set about, single-handed, to build a 'live aboard' forty-foot steel yacht. Nine hard months later, he launched 'Scotia.' She underwent her sea trials during the 1997 Banff to Stravanger yacht race.

At the turn of the century, Hugh's dreams came true and with Brenda, sailed away into the wide blue yonder.

Since then they have sailed around the world, going north to Norway, west across the Atlantic and the Pacific and South to New Zealand. They returned to the Mediterranean by way of the South China Sea, and the Suez Canal, in 2010.

Other paperback books by Hugh Fraser from **Lulu Publishing**.

A Gift from a Princess. ISBN 978-1-4477-3062-0
a seafaring story for those who dream of adventure

Stand and Run. ISBN 978-1-4477-4716-1
An amusing adult thriller.

All formats of **e-books** are available from <**Smashwords.com**>
Loch Ness Tales Legends and Recipes. __ISBN-1-4658-6981-4
A Gift from a Princess. _____ISBN-1-4660-1718-4
Stand and Run._____ISBN-1-4661-5968-6
Also by the major online retailers such as:- Apple, Barnes & Noble, Sony, Kobo, Borders Australia and Angus & Robertson Australia (both powered by Kobo), Whitcoulls (New Zealand, powered by Kobo), the Diesel eBook Store and ScrollMotion
Also at Amazon Kindle note, no ISBN Numbers. #

www.ingramcontent.com/pod-product-compliance
Lightning Source LLC
Chambersburg PA
CBHW032008170526

45157CB00002B/595

* 9 7 8 1 4 4 7 7 2 9 5 6 3 *